PROCESSING OF HIGH-TEMPERATURE SUPERCONDUCTORS AT HIGH STRAIN RATES

HOW TO ORDER THIS BOOK

BY PHONE: 800-233-9936 or 717-291-5609, 8AM–5PM Eastern Time

BY FAX: 717-295-4538

BY MAIL: Order Department
Technomic Publishing Company, Inc.
851 New Holland Avenue, Box 3535
Lancaster, PA 17604, U.S.A.

BY CREDIT CARD: American Express, VISA, MasterCard

BY WWW SITE: http://www.techpub.com

Processing of High-Temperature Superconductors at High Strain Rates

A. G. Mamalis, Dr.-Ing.,
Academician Professor

D. E. Manolakos, Dr.-Ing.,
Associate Professor

A. Szalay, Dipl.-Ing.

G. Pantazopoulos, Dr.-Ing.

TECHNOMIC PUBLISHING CO., INC.
LANCASTER • BASEL

Processing of High-Temperature Superconductors at High Strain Rates

a **TECHNOMIC**® publication

Technomic Publishing Company, Inc.
851 New Holland Avenue, Box 3535
Lancaster, Pennsylvania 17604 U.S.A.

Printed in the United States of America
10 9 8 7 6 5 4 3 2 1

Main entry under title:
Processing of High-Temperature Superconductors at High Strain Rates

A Technomic Publishing Company book
Bibliography: p.
Includes index p. 259

Library of Congress Catalog Card No. 00-102582
ISBN No. 1-56676-878-0

CONTENTS

FOREWORD

The discovery of superconductivity at high temperature in 1986 has set fire to the research on cuprates. Thousands of articles have been published these last 10 years on high-T_c superconducting cuprates. In a similar way, many books have been written on the physics of these compounds. In contrast, only few lines have been written on the applied side of these materials.

At this point of the investigations, it is time to take stock about the processing of these materials in view of their applications. There is no doubt that several of these numerous superconductors, such as the famous "123" YBCO or the Bi "2212" or "2223" cuprates will be used in devices in the next years.

Specialists of the processing of these materials, A. G. Mamalis and his coworkers and A. Szalay, have performed a very efficient work in this field. Through this book, the reader benefits from the high experience of the authors in the field of the processing of bulk superconductors at high strain rates. Not only do the authors describe conventional methods of fabrication of bulk superconductors and thin films, but they also consider the impact loading of solid and porous media from the theoretical and experimental viewpoint.

Besides these methods, which are of capital importance for the elaboration of optimized high-T_c superconductors, this book relates the main methods of characterization of these materials and gives an excellent overview of their field of applications.

This book is well structured and conceived in such a way that it can be used by beginners in the elaboration of materials and by specialists who need details and references in this area. There is no doubt that this work will be used as a reference by students and scientists who are interested in the processing of HTS.

BERNARD RAVEAU
Caen

PREFACE

The introduction that follows sufficiently describes our aims in presenting this book on the net-shape processing of high-temperature advanced superconducting ceramics, which are subjected to elevated, and in particular, high-strain rates, relevant to design and manufacture of electrical/electronic, transportation and bioengineering equipment. Essentially, it comprises the results of our extensive theoretical and experimental work on the topic and the material of a series of lectures on net-shape manufacturing of advanced materials given to undergraduates and graduates in Mechanical, Material Science and Electrical/Electronic Engineering. The book is intended to illustrate and indicate the engineering design outlets and applications of the analytical and experimental work, mainly on macro- and micromechanics, static and dynamic powder consolidation and processing, stress-wave propagation into solid and porous materials, as well as applications of bulk and thin-film superconductors and the new notions and considerations in applied engineering applications. We hope that the contents of our work will be of value to students, teachers and many kinds of professional engineers.

A. G. MAMALIS
G. PANTAZOPOULOS
A. SZALAY
D.E. MANOLAKOS

CHAPTER 1

Introduction

1.1 BRIEF HISTORY OF SUPERCONDUCTIVITY

The historical evolution of the superconducting materials is presented diagramatically in Figure 1.1.

On July 10, 1908, Heike Kamerlingh Onnes at the University of Leiden in the Netherlands successfully operated the first liquefier of helium, which exhibited a boiling point of approximately 4.2 K at atmospheric pressure. Since then, Onnes and other researchers tried to determine the electrical properties of matter at very low temperatures. In some metallic materials, such as copper (Cu), platinum (Pt), gold (Au), showing a continuous decrease of electrical resistance with decreasing temperature, a finite value of resistance at liquid helium temperature was finally obtained. Only mercury (Hg) revealed a rapid decrease in the resistance at 4.25 K, which was vanished at about 4.2 K (boiling point of He), see Figure 1.2. The mercury-work was first reported by Onnes in the 1st Solvay Congress in Brussels, during October 1891 and received the Nobel Prize in Physics in 1913 [1].

Onnes followed this work showing that indium (In), tin (Sn) and lead (Pb) became superconducting at 3.4, 3.7 and 7.2 K, respectively. These materials, that present zero resistance at a certain temperature above absolute zero, are named *superconductors* and the related phenomenon *superconductivity.* A generalized comparison of resistivity for normal metal and superconductors as a function of temperature is shown in Figure 1.3.

The highest *critical temperature* of pure superconductors was shown in the case of niobium (Nb), possessing superconductivity below 9.5 K. Note that the research in superconductivity has led to the formation of superconducting alloys and compounds based on Nb. Characteristic Nb-alloy systems are the Nb-Ti and the Nb-Zr, but the highest critical temperatures, i.e., 18 and 23 K,

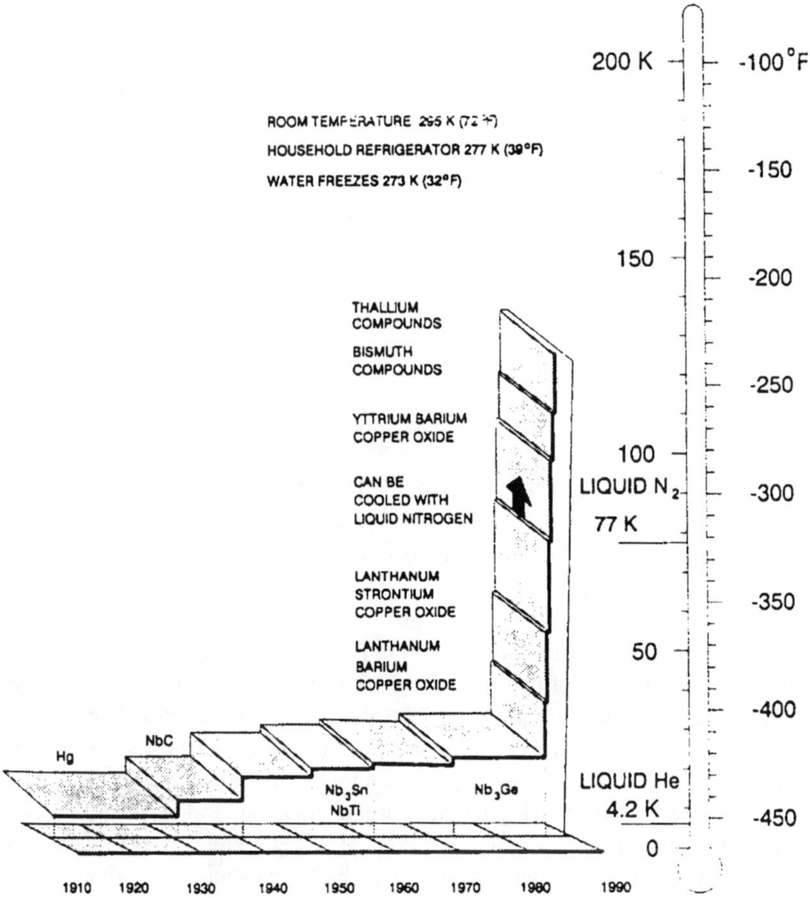

FIGURE 1.1. Evolution of superconductors.

are indicated by the intermetallic compounds Nb_3Sn and Nb_3Ge, respectively, which possess a characteristic β-W type lattice (A_3B).

In 1986 a *high*-T_c or *high-temperature superconductivity* was invented by Bednorz and Müller at IBM in Zürich [2]. The new superconductor was a mixed oxide of lanthanum, barium and copper (La-Ba-Cu-O) of perovskitic crystal structure; the critical temperature of this ceramic material was higher than 30 K. It has to be noted, that, after several false starts and disappointments, the turning point for high-T_c superconductivity come in late 1985 when Bednorz was alerted to a publication by the French team of Michel and Raveau in Caen [3], who reported metallic-like conductivity in La-Ba-Cu-O in the range from −150°C to 300°C. The French team had not been investigating superconduc-

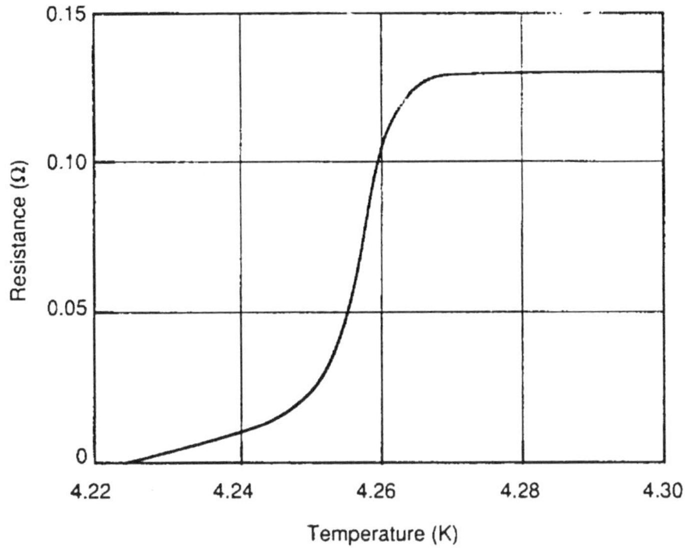

FIGURE 1.2. Resistivity-temperature diagram for mercury (Hg).

tivity at that time but had shown several years before, see Reference [4], the possibility to stabilize the mixed valent state copper at normal pressure, in 2D cuprate in connection with the Jahn Teller effect of copper. Bednorz and Müller received the Nobel Prize in Physics, for their invention, in 1987.

The main high-T_c superconductors, from a practical application point of view, were invented afterward. Chu and Wu and their group at Texas Center of High-T_c Superconductivity (TCSUH) in Houston, Texas, discovered the superconducting ceramic $YBa_2Cu_3O_{7-x}$ (1-2-3 material) with critical tempera-

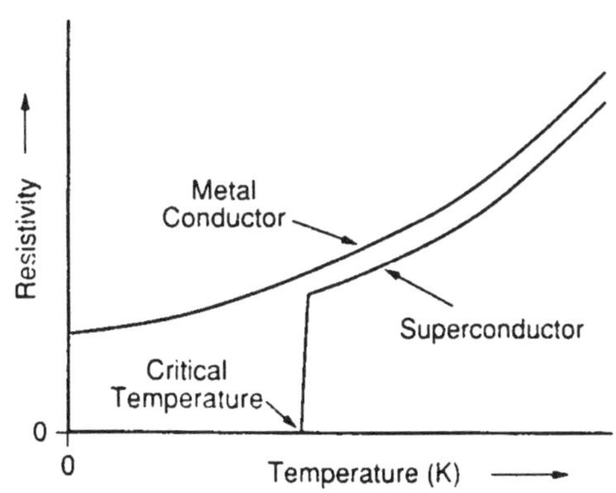

FIGURE 1.3. Resistivity-temperature diagram for a superconductor and a metal-conductor.

ture up to 93 K above liquid nitrogen temperature [5]. This fact makes the use of superconductivity very profitable because liquid nitrogen as the cooling medium is much cheaper and more easily handled than helium.

Maeda's group at Tsukuba Labs in Japan synthesized the Bi-Sr-Ca-Cu-O oxide, becoming superconductor at 110 K [6]. More ductile and stable than the orthorombic 1-2-3 compound, it exhibits several superconductive or normal phases that are not trivial to be separated.

Another characteristic Cu-O superconductor is the Tl-Ba-Ca-Cu-O, invented by Sheng and Hermann and their group from the University of Arkansas [7]. Similarly to Bi-Sr-Ca-Cu-O oxide, the Tl-based material is a multiphase system; the highest critical temperature possessed is approaching 125 K.

1.2 THE STRUCTURE OF THE MONOGRAPH

The discovery of high-temperature superconductivity by Bednorz and Müller in the La-Ba-Cu-O system resulted in very extensive research work about the discovery and synthesis of other high-temperature superconductors, such as Y-Ba-Cu-O and Bi-Sr-Ca-Cu-O. These new superconducting materials, possessing superconductivity above liquid nitrogen boiling point, are used in many engineering applications, from electronic sensors to rotating electrical generators and from nanometer-scale thin films to kilometer-long wires and coils. Therefore, design and net-shape manufacturing of superconducting components, starting from the initial synthesized powders, is nowadays of utmost industrial importance.

This book is primarily focused on the bulk-fabrication techniques of high-temperature ceramic superconducting components, especially on the combination of dynamic powder-consolidation and subsequent deformation processing. The properties of these ceramics, which are difficult-to-be-formed materials by applying conventional techniques, are combined for the net-shape manufacturing of such components for the construction of HTS devices; this is the core of the superconductor research work that has taken place through the collaboration of the Manufacturing Technology Division of the National Technical University of Athens, Greece, and the Metalltech Ltd. of Budapest, Hungary, according to the fabrication sequence, from the raw material to industrial applications as shown in Figure 1.4. However, very important topics, such as superconducting structures, chemical synthesis, film fabrication and characterization techniques, are also reviewed throughout this book, to provide a complete comprehensive view of superconductors engineering.

In Chapter 2 the fundamental aspects of superconductivity are outlined, whereas Chapter 3 deals with the most important chemical synthesis methods of most of the high-temperature superconducting compounds.

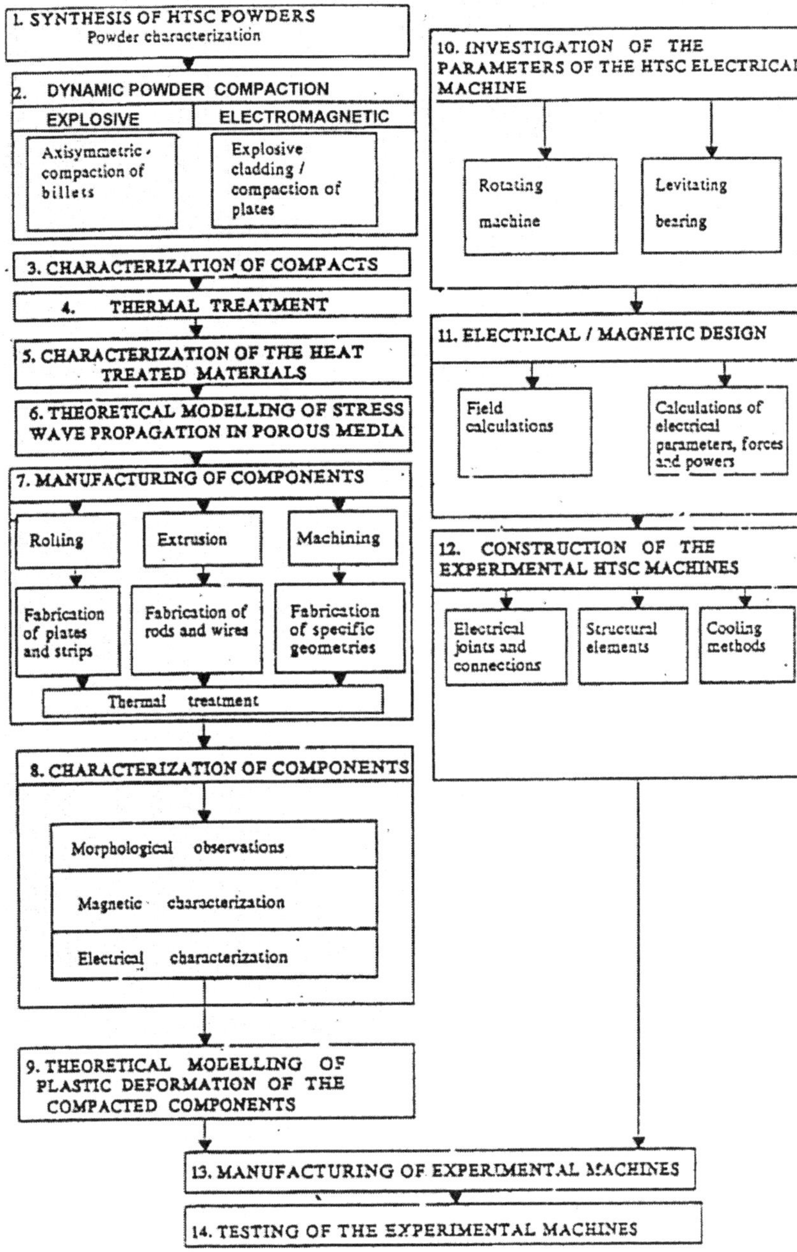

FIGURE 1.4. Process flow diagram from the powder synthesis to the final stages of fabrication and applications of high-T_c superconductors.

In Chapter 4 the dynamic powder compaction techniques and extensive experimental pertaining to the effect of shock waves on microstructure, properties and defect formation of impacted superconductive components of the YBCO and BSCCO compounds are discussed in detail. In addition, the theoretical modeling of stress wave propagation through the porous media, resulting in the consolidation of the powders, is attempted, considering also the Hugoniot curves and equation-of-state (EOS) of the materials.

The deformation processing of HTS compacts in various geometries in plastic deformation-microstructure-processing-properties relationships, is presented in detail in Chapter 5. The defects and the deformation mechanisms for the forming techniques used are also considered. Furthermore, other bulk-processing techniques, such as melt-texturing, composite reaction texturing and mechanical texturing, are briefly discussed.

In Chapter 6 the processes used for the fabrication of HTS films are discussed in detail.

Chapter 7 deals with the physicochemical techniques used for the characterization of HTS powders and components, whereas in Chapter 8 the most important HTS applications are presented. Finally, in Chapter 9 some new classes of superconducting materials and their applications are suggested, whereas future perspectives of high-T_c superconductivity and its techniques toward higher critical temperatures are outlined.

This book is addressed to scientists, researchers and engineers involved in the multidisciplinary field of high-temperature superconductivity approaching the subject more from an engineering point of view. It is our hope that the contents of this monograph will be of value to students, teachers and many kinds of professional engineers and, thus, enhance interest in superconducting materials and the related technology.

1.3 REFERENCES

1. Onnes H. K. (1911), "On the sudden change in the rate at which the resistance of mercury disappears," Communication from the Physical Laboratory of the University of Leiden, No. 124c.
2. Bednorz J. G. and Müller K. A. (1986), "Possible high-T_c superconductivity in the Ba-La-Cu-O system," *Z. Phys.* B64, 189.
3. Michel C. and Raveau B. (1984), "Oxygen intercalcination in mixed valence copper oxides related to the perovskite," *Revue de Chimie Minérale* 21, 407.
4. Nguyen N., Choisnet J., Hervieu M. and Raveau B. (1981), "Oxygen defect K_2NiF_4-type oxides: the compounds $La_{2-x}Sr_xCuO_{4-x/2+\delta}$," *J. Solid State Chem.* 39, 120.
5. Wu M. K., Ashburn J. R., Tang C. J., Meng R. L., Ga L., Huang Z. J., Wang Y. Q. and Chu C. W. (1987), "Superconductivity at 93 K in a new mixed phase Y-Ba-Cu-O compound system at ambient pressure," *Phys. Rev. Lett.* 58, 908.

6. Maeda H., Tanaka Y., Fukutomi M. and Asano T. (1988), "A new high-T_c oxide superconductor without a rare earth element," *Jap. J. Appl. Phys. Lett.* 27, 209.
7. Sheng Z. Z., Kiehl W., Bennet J., El Ali A., Marsh D., Mooney G. D., Arammash F., Smith J., Viar D. and Hermann A. M. (1988), "New 120 K Tl-Ba-Ca-Cu-O superconductor," *Appl. Phys. Lett.* 52, 1738.

CHAPTER 2

Fundamentals of Superconductors

2.1 NOTATION

B = magnetic induction
c = heat capacity
e = charge of electron
F_L = Laplace force
h = Planck's constant
H = magnetic field intensity
H_c = critical magnetic field intensity
H_{c1} = lower critical field
H_{c2} = upper critical field
H_o = critical field intensity at zero Kelvin temperature
I = supercurrent flow
J = current density
J_c = critical current density
k_B = Boltzmann's constant
M = magnetization
n = integer
$N(E_F)$ = energy state density below Fermi level
T = temperature
T_c = critical temperature
U = voltage
V = correlation factor
α = fitting constant
Δ = semiwidth of energy gap
κ = parameter ($= \lambda/\xi$)
λ = penetration depth (London) of magnetic field

M = isotopic mass
ξ = coherence length
ρ_o = resistive transition
Φ = magnetic flux
Φ_o = flux quantum
$\psi(x)$ = wave function
ω = frequency
ω_D = Debye's frequency

2.2 BCS THEORY

The explanation of the superconductivity theory, mainly in metallic systems, originated from the American researchers Bardeen, Cooper and Schrieffer in 1957 [1], who won the Nobel Prize in 1972. The Schrieffer corresponding theory is abbreviated as *BCS theory.*

The basic characteristic effects of the superconductive state is the persistence of the electric currents, the diamagnetic effect and the specific heat discontinuity. From electrical measurements, the elementary charge of a material in the superconducting state was found equal to $2e$, where e is the charge of the electron (1.6×10^{-19} Cb). This result yields to the conclusion that, the current flowing in a superconducting material is composed of electron pairs, which are called *Cooper pairs,* and this pairing occurs despite the repulsive interaction between the two electrons. This attraction may be simply explained by the "mattress effect," i.e., when a heavy ball rolls fast on a soft mattress, the mattress will bend downward or sink where the ball is, whereas in the case of the ball rolling very fast, the time for back relaxation is not enough for the springs to return to the starting position immediately after the passage of the ball. If, now, a second ball is traveling on the same mattress and comes closer to the first ball, the mattress will sink, bringing the two balls together. Substituting electrons for balls and the solid composed of sluggish ions for the mattress, an attractive interaction between electrons is, thus, obtained.

The electron-pairing mechanism is based on the so-called *electron-phonon interaction.* An electron moving in a solid lattice can be affected by another electron via acoustic quanta (i.e., phonons), which are originated from the vibrations of the lattice atoms. The electron-electron interactions are realized through phonon exchange, leading to electron condensation, being responsible for the neutralization of Coulomb repulsion. The electronic pair consists of two electrons possessing opposite momentum and spin, and the spacing between them is known as coherence length ξ. The coherence length is of the order of $100 \div 1000$ nm in pure superconducting materials. At temperatures higher than the critical temperature, the thermal energy obtained results in breaking the Cooper pairs and stopping the superconduction. The BCS theory may be, finally, summarized as

$$T_c = \frac{1.14h\omega_D \exp[-1/N(E_F)V]}{2\pi k_B} \tag{2.1}$$

where, following the Notation section, h is the Planck's constant, ω_D the Debye's frequency, $N(E_F)$ the energy state density below Fermi level, V a correlation factor and k_B the Boltzmann's constant.

2.3 BASIC NOTIONS

The fundamental phenomena related to the superconductivity may be listed as [2]:

- zero resistivity
- diamagnetic or Meissner effect
- energy gap
- specific heat discontinuity
- magnetic flux quantization
- isotopic effect
- tunneling effect

2.3.1 Zero Resistivity

The curve representing the variation of resistivity as a function of temperature is shown in Figure 2.1. By convention, the *onset temperature,* $T_{c,\text{onset}}$, is defined as the temperature value at which the slope changes entering from the normal to the transition behavior and the *critical temperature,* T_c, as the temperature corresponding to *zero resistivity point.*

Measurements of resistivity or magnetic susceptibility may lead to the determination of T_c. For narrow transitions, the transition temperature corresponds to the zero resistivity. However, because the resistivity slope near T_c may widely vary between ac and dc measurements, a corresponding difference in the measured value of T_c may be shown. Long resistive tails indicating the presence of secondary phases, superconducting or not, make the determination of T_c difficult. By convention, T_c is found with respect to the half of the resistive transition corresponding to $\rho_o/2$, see Figure 2.1. T_{co} is the onset and T_{cf} is the offset of the transition (end or zero resistivity point).

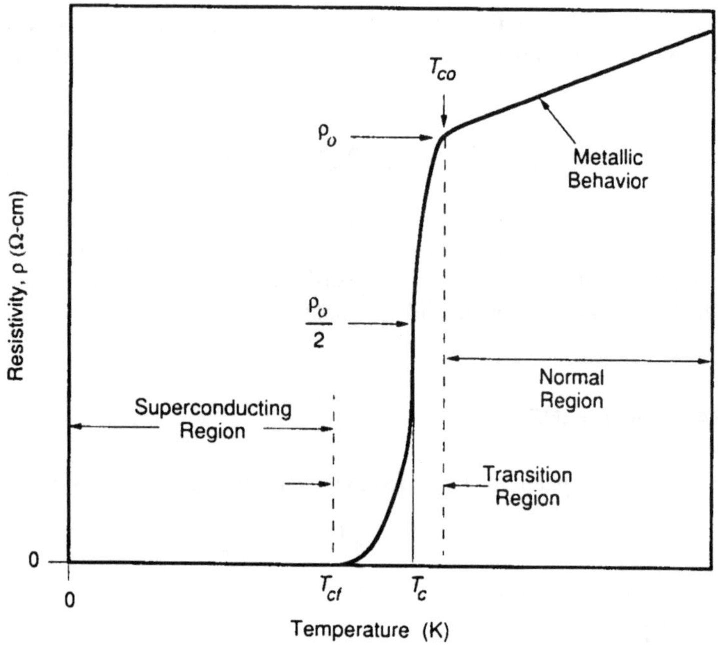

FIGURE 2.1. Characteristic resistivity-temperature curve for a superconductor.

2.3.2 Diamagnetic or Meissner Effect

Apart from the outstanding properties, the superconductive materials possess exceptional magnetic properties. When a magnetic field of intensity H is applied to a superconductive material ($T < T_c$), exclusion of magnetic field lines from the interior of the material has been observed. This is valid when the field intensity H is lower than a critical intensity H_c. The ideal superconducting material behavior is then called *pure diamagnetic behavior,* while the magnetic susceptibility is equal to -1. This phenomenon is known as *Meissner-Ochsenfeld* or *diamagnetic effect.*

For temperatures higher than T_c, the diamagnetic effect is eliminated, and the magnetic field penetrates in the interior of the material.

A schematic representation of the diamagnetic effect is illustrated in Figure 2.2(a), while the variation of H_c as a function of temperature is shown in Figure 2.2(b), given by the equation

$$H_c = H_o\left[1 - \left(\frac{T}{T_c}\right)^2\right] \tag{2.2}$$

where H_o is the critical intensity at zero Kelvin temperature.

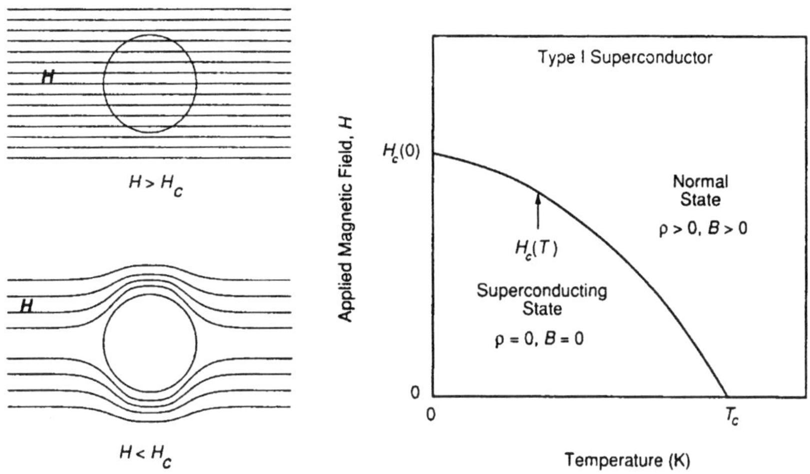

FIGURE 2.2. (a) Schematic representation of the Meissner-Ochsenfeld effect; (b) evolution of the critical magnetic field as a function of temperature for a Type I superconductor.

The magnetization, M, defined by the relation (see also Notation section) $B = \mu_o(M + H)$, for a certain temperature $T < T_c$ varies with magnetic field according to the curve of Figure 2.3(a).

LONDON EQUATION—PENETRATION DEPTH

The electrodynamic behavior of a superconducting material is described by London's equations. By combining with Maxwell's equation a fundamental relation is obtained

$$\nabla^2 H = \frac{H}{\lambda^2} \tag{2.3}$$

where λ is the characteristic penetration depth (London) of the magnetic field in the interior of the material.

Under the semi-infinite solid assumption, the solution of the differential equation yields the following spatial distribution of the magnetic field

$$H(x) = H(0) \cdot \exp(-x/\lambda) \tag{2.4}$$

The value of penetration depth is of the order of magnitude of 50 nm for pure metals superconductors.

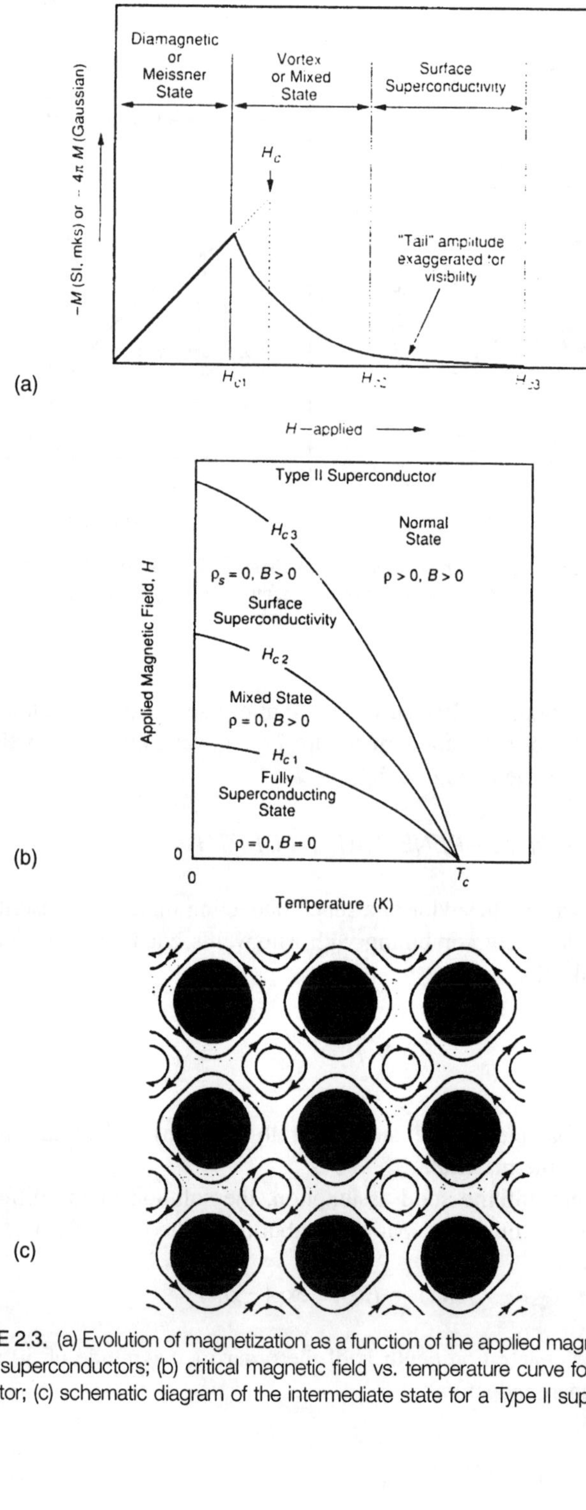

FIGURE 2.3. (a) Evolution of magnetization as a function of the applied magnetic field for Type I and II superconductors; (b) critical magnetic field vs. temperature curve for a Type II superconductor; (c) schematic diagram of the intermediate state for a Type II superconductor.

GINSBURG-LANDAU THEORY

The phenomenology of the superconductivity has been thoroughly studied by Ginsburg-Landau in 1957. Assuming that the characteristic wave function $\psi(x)$ is describing the electronic fluid state, then the expression $[\psi(x)]^2$ is a measure of the density of the superconductive electrons. The wave function ψ in the case of the Ginsburg-Landau superconductivity theory is called *order parameter*.

Ginsburg and Landau have also introduced the κ-parameter, which represents the ratio of the characteristic lengths λ and ξ, i.e.,

$$\kappa = \frac{\lambda}{\xi} \tag{2.5}$$

TYPES OF SUPERCONDUCTORS

Type I Superconductors ($\kappa < 1/\sqrt{2}$)

This type of superconductor possesses all the diamagnetic characteristics that have been already mentioned. All pure metals, except niobium (Nb), belong in this category.

Type II Superconductors ($\kappa > 1/\sqrt{2}$)

In this class of superconductors two different values of the critical magnetic field, a lower critical field H_{c1} and an upper critical field H_{c2}, are present. When the applied field $H < H_{c1}$, then the diamagnetic effect (exclusion of magnetic field lines from the interior of the material) is observed. In the case of $H_{c1} < H < H_{c2}$, progressive penetration of the magnetic field begins forming an intermediate or mixed or vortex state. Finally, when H exceeds H_{c2}, the magnetic field totally penetrates into the interior of the material, destroying, therefore, the superconducting state in the interior of the material that remains in a surface superconducting state for $H_{c2} < H_c < H_{c3}$, see Figure 2.3(b). Type II superconductors are all the alloys (including Nb), intermetallic components and ceramic oxides.

INTERMEDIATE STATE

The *intermediate state,* denoted also as *mixed state, vortex state* or *Schubnicov phase,* appears in Type II superconductors when the applied magnetic field H lies between the values of lower and upper critical field, H_{c1} and H_{c2}, respectively. In this case, the progressive penetration of magnetic field lines into the interior of the material forms normal, nonsuperconducting regions, adjacent to superconducting ones. The field penetration creates, therefore, the so-called vortices or fluxoids in an ordered array, which is known as *Abrikosov lattice,* see Figure 2.3(c). Every fluxoid is a closed loop characterized by a supercurrent flow I. The increase of the external field gives rise to the magnetic induction, B, in the vortex interior causing Laplace forces ($F_L \propto B \cdot I$) tending to move the vortices (characterized as flux flow or flux creep). The spreading of the vortices results in the decrease or even the total elimination of the superconducting phase.

FLUX PINNING

Considering that the flow of the electric current causes magnetic field (Ampere's law), every superconductor of Type I may hold electric current, having a density lower than a critical one, J_c. Above this *critical current density* J_c the induced field exceeds the critical field value, H_c resulting in the destruction of the superconducting state.

For the Type II superconductors the critical density, J_c corresponds to the current value at which the vortex flow starts to occur. The degree of difficulty of the fluxoid motion defines the ability of the material to stay at the diamagnetic (superconductive) state. This characteristic material property is known as *flux pinning*. Increased flux pinning results in high values of J_c. Therefore, vortex stability and J_c can increase by the introduction of impurities or defects that act as flux trappers, stabilizing, therefore, the vortices. It is important that this array of fluxoids remain stable, because their motion leads to energy dissipation by normal currents induced in the vortex cores.

Metallurgical defects, including dislocations, point defects, precipitates, secondary phases and grain boundaries, resulted from processing techniques or from radiation damage, for example neutron irradation, may serve to pin the vortices found in Type II materials. Chemical doping results also in enhancement of flux pinning in novel oxide superconductors [3]. Effective pinning is achieved when the pinning defects and their intermediate spacings are of the order $(1 \div 5)\xi$, where ξ is the coherence length.

It is noteworthy that, although T_c and H_c may be generally considered as intrinsic material properties, J_c is a function of microstructure and can be varied several orders of magnitude, depending on the material processing technique.

2.3.3 Energy Gap

During the normal to superconducting transition, the electronic structure is reconstructed to allow the developed electron pairs to construct the superconducting current. This new electronic structure results in an anomaly in the continuity of the permitted energy levels, creating, therefore, near the Fermi surface, a forbidden *energy gap* of width equal to 2Δ. When the energy gap is exceeded, the breaking of electron pairs takes place, leading to the destruction of the superconductivity. Taking into account the BCS theory, the width of the energy gap 2Δ increases with increasing critical temperature according to

$$2\Delta = 3.528k_BT_c \tag{2.6}$$

where k_B is the Bolzmann's constant.

2.3.4 Specific Heat Discontinuity

An abrupt change of heat capacity c is observed during the normal to superconductive transition, which may be explained by the presence of the energy gap. The change of the specific heat as a function of temperature is presented in Figure 2.4.

2.3.5 Magnetic Flux Quantization

The magnetic flux, passing through a superconductive ring, is an integer multiple of the flux quantum Φ_o, passing through a single fluxoid, i.e.,

$$\Phi = n\Phi_o = n\frac{h}{2e} \tag{2.7}$$

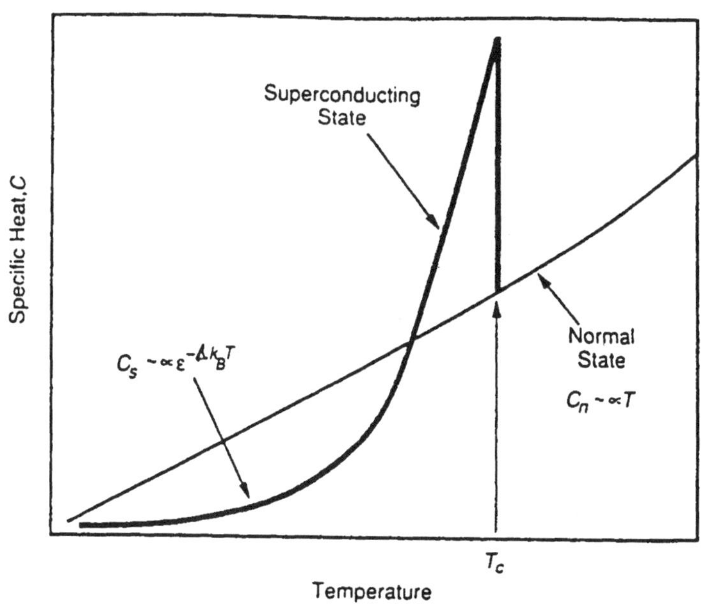

FIGURE 2.4. Evolution of specific heat as a function of temperature (from Reference [2]).

where n is an integer number, h the Planck's constant and $2e$ the elementary charge of the superconductive material.

2.3.6 Isotopic Effect

The critical temperature of a superconducting material varies with the isotopic mass, M. The critical temperature changes smoothly by mixing several isotopes of the same element. The experimental results of an isotope series can be summarized by the following relationship

$$M^{\alpha} T_c = \text{const.} \tag{2.8}$$

where α is a fitting constant.

Constant α is close to 0.5 for pure metals superconductors. This can be explained by the BCS theory where the T_c is proportional to the Debye's frequency, $\omega_D \sim M^{-1/2}$. The relation $\omega_D \sim M^{-1/2}$ is valid under the assumption that the lattice atoms behave as ideal harmonic oscillators.

2.3.7 Tunneling Effect (Josephson Effect)

The *tunneling* or *Josephson effect* constitutes one of the most important superconductivity-based phenomena. Microelectronic applications and measuring equipment of very high degree of resolution, such as SQUIDs (Superconductive Quantum Interference Devices), are based on this effect.

In the case of the Josephson effect, a normal insulating interface with a

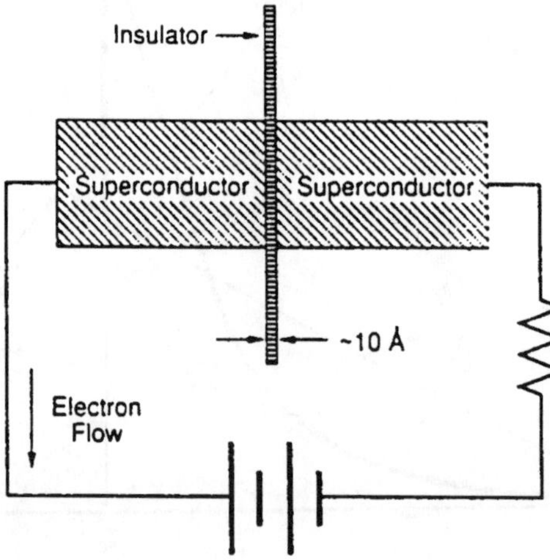

FIGURE 2.5. Characteristic Josephson sandwich material structure (from Reference [2]).

thickness 1 ÷ 5 nm is placed between two superconductive components, see Figure 2.5. When a direct current voltage U is applied, then an alternating current is flowing in S-I-S sandwich structure, with a constant frequency ω of magnitude

$$\omega = \frac{2\pi(2e)U}{h} \tag{2.9}$$

where, following the Notation section, h is the Planck's constant and $2e$ the elementary charge of the superconductive material.

2.4 REFERENCES

1. Bardeen J., Cooper L. N. and Schrieffer J. R. (1957), "Theory of superconductivity," *Phys. Rev.* 108, 1175.
2. Doss J. D. (1989), *Engineer's Guide to High-Temperature Superconductivity,* John Wiley & Sons, New York.
3. Shimoyama J., Kitazawa K. and Kishio K. (1996), "Strong flux pinning, anisotropy and microstructure of (Hg, Re)M2Can-1CunOy (M = Ba,Sn)," *Proc. 10th Anniversary HTSC Workshop on Physics, Materials and Applications,* Houston, Texas, 85.

CHAPTER 3

Synthesis of Ceramic Superconductors

3.1 GENERAL

In the beginning of this chapter, which deals with the basic chemical processes pertaining to the fabrication of high-T_c ceramics, an attempt is made to classify and describe the most important families of the superconducting cuprates from their crystal structure point of Y-Ba-Cu-O, Bi-Sr-Ca-Cu-O, Tl-Ba-Ca-Cu-O and Hg-Ba-Ca-Cu-O.

Furthermore, the principal synthetic methodologies for obtaining high-T_c superconducting cuprates, starting from the traditional solid-state process (ceramic method) to more nonconventional techniques, such as the coprecipitation/precursor synthesis, the sol-gel technique and the combustion synthesis (SHS), are reported and discussed with the involved operational parameters, precautions and materials to be produced. The fabrication methods described are dealt with the synthesis of superconducting powders from the initial raw materials, which is the primary production stage in the sequence of the manufacture of functional superconducting components (bulks or films) for various engineering applications; see the process sequence in Figure 1.4.

Finally, the possible alterations in the initial raw materials chemistry, which may enhance the expected superconducting properties, such as the current density, J_c, and the critical temperature, T_c, are examined. Certain elemental additions (doping) may affect thermodynamic and/or kinetic factors of the formation of the superconducting phase or even lead to the production of new interesting phases. Note that the presence of extrinsic dopant may sometimes help the creation of superconducting solid solutions, i.e., for example, the partial substitution of Ba by K in the Y-Ba-Cu-O lattice, with higher critical temperature.

3.2 CLASSIFICATION OF HIGH-T_c SUPERCONDUCTING COMPOUNDS (HTS)

The nature and the superconducting properties of the main high-T_c cuprate systems, namely the La-Ba-Cu-O, Y-Ba-Cu-O, Bi-Sr-Ca-Cu-O, Tl-Ba-Ca-Cu-O and Hg-Ba-Ca-Cu-O, are analyzed by outlining the crystal features of the main members of the homologous series of the above-mentioned groups.

3.2.1 La-Ba-Cu-O Cuprate System

In 1986, Bednorz and Müller reported their discovery on a new nonmetallic superconductor that possesses a transition at a temperature of 35 K [1], i.e., 12 K higher than the transition temperature of Nb_3Ge compound, and won the Nobel Prize in Physics. The new superconducting material was an oxide of the La-Ba-Cu-O family, i.e., $La_{2-x}Ba_xCuO_4$ ($x = 0.2$), with a K_2NiF_4 crystal structure that is perovskite-related layer compound.

The perovskite structure is named according to the primary found mineral $CaTiO_3$, and it is a very common structure for natural minerals and also for industrial ceramics [2,3]. Many compounds of the general formula ABX_3 (X: O, F, S) adopt this type of structure. The so-called *A-type perovskite,* is presented in Figure 3.1(a), where the central cation A (Ca) is coordinated by eight cations B (Ti), positioned at the corners of the cube, and by 12 anions X (O) at the midpoint of the edges. Considering the B-cation site as the center of the unit cell a *B-type perovskite* structure is yielded, see Figure 3.1(b). The K_2NiF_4 and, therefore, the $La_{2-x}Ba_xCuO_4$ crystal structure, can be divided into three subunits, see Figure 3.2: the central unit which is a B-type perovskite structure is sandwiched by the two (top and bottom) A-type perovskite

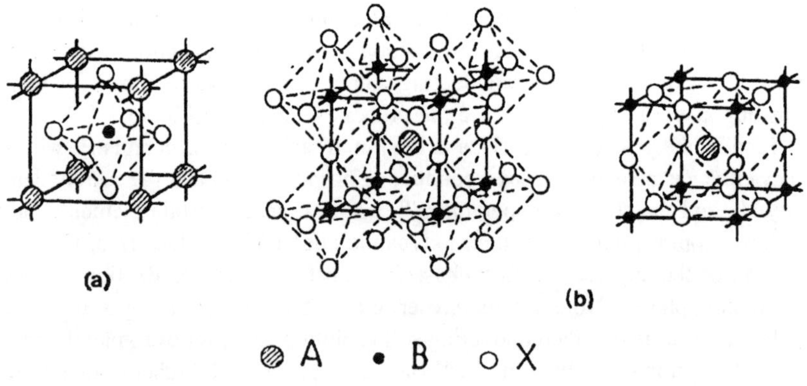

FIGURE 3.1. (a) B-type and (b) A-type perovskite crystal structures.

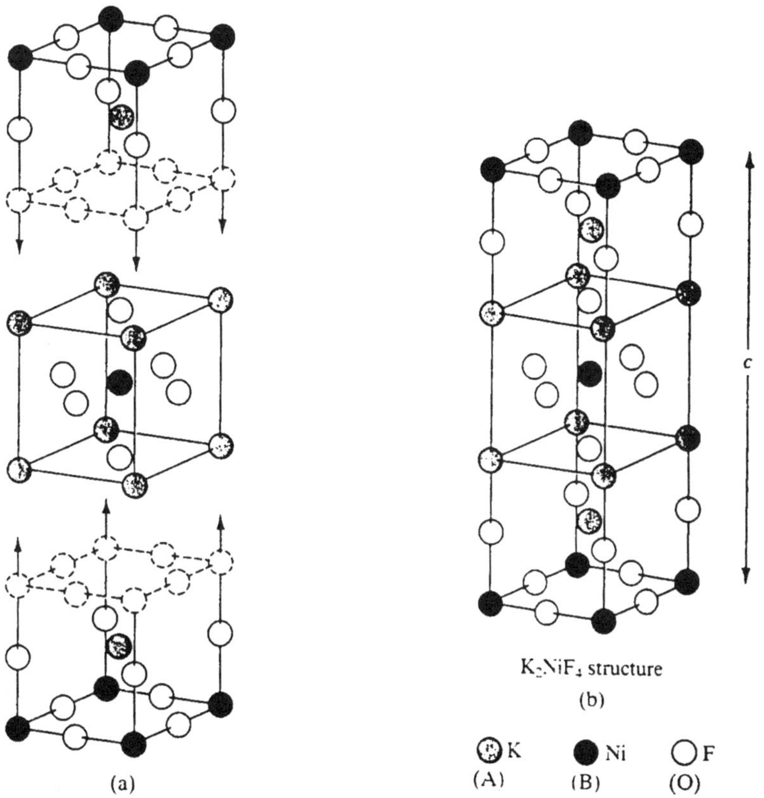

FIGURE 3.2. K_2NiF_4 ($La_{2-x}Ba_xCuO_4$) perovskite-related layered crystal structure.

cells missing the dotted shown atomic layers. The La and Ba atoms are very close in their dimensions and, therefore, they are distributed into the A (K) lattice sites, whereas Cu atoms take the B (Ni) sites.

The unit cell of $La_{2-x}Ba_xCuO_4$ is characterized by a tetragonal crystal structure with lattice dimensions $a = b = 0.378$ nm and $c = 1.32$ nm. Extensive reviews on perovskitic superconductive crystals have been presented by Raveau et al. [4] and Aleksandrov and Beznosikov [5].

3.2.2 Y-Ba-Cu-O Cuprate System

The high-temperature ceramic superconductor, namely $YBa_2Cu_3O_{7-\delta}$, discovered by Wu and Chu, has a maximum T_c up to 93 K, which is higher than the boiling point of nitrogen [6]. The structure of this oxide is based on triple perovskite unit cell of the $Y_3Cu_3O_9$ compound, see Figure 3.3(a). If two of

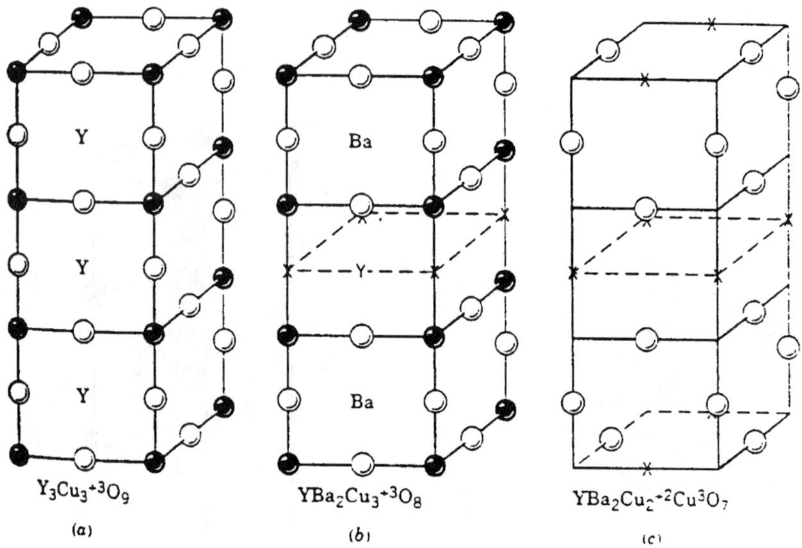

FIGURE 3.3. Origin of the $YBa_2Cu_3O_{7-\delta}$ triple-perovskite crystal structure.

the three ions Y^{3+} are substituted by Ba^{2+} ions, then the $YBa_2Cu_3O_8$ structure is formed, see Figure 3.3(b). For electroneutrality purposes, the removal of one oxygen anion (O^{2-}) is necessary, accommodated by forming vacancies around the central yttrium ion. The main reason of the existence of superconductivity in this cuprate is the presence of trivalent copper ions (Cu^{3+}). In the 123-YBCO compound, almost one third of copper ions are in the +3 valence state, reducing further the oxygen content and resulting in the $YBa_2Cu_3O_{7-\delta}$ structure. The additional vacancies are formed at the top and bottom of the Cu planes, resulting in a structure illustrated in Figure 3.3(c) in an abstracted form.

From this new type of crystal structure, two nonequivalent types of copper atoms are developed: (i) the *Cu(1) atoms,* oriented along the [100] crystallographic directions, which are coordinated with four oxygen atoms, forming the so-called *Cu-O chains,* and (ii) the *Cu(2) atoms,* lying on the [001] *Cu-O planes* (or *ab* planes), which are coordinated with five oxygen atoms. The fivefold coordinated sites form the bases of the square pyramids (tetrahedra) joined at their corners, see Figure 3.4.

The oxygen stoichiometry, which depends on the copper valence state, is very sensitive to the temperature and the oxygen pressure and, it varies between 6 and 7 ($\delta = 0 \div 1$). When $\delta = 0 \div 0.5$ the YBCO compound has an *orthorhombic crystal structure* that possesses superconductivity. The ideal superconducting properties are exhibited when $\delta = 0.08$ ($O_{6.92}$) with the max-

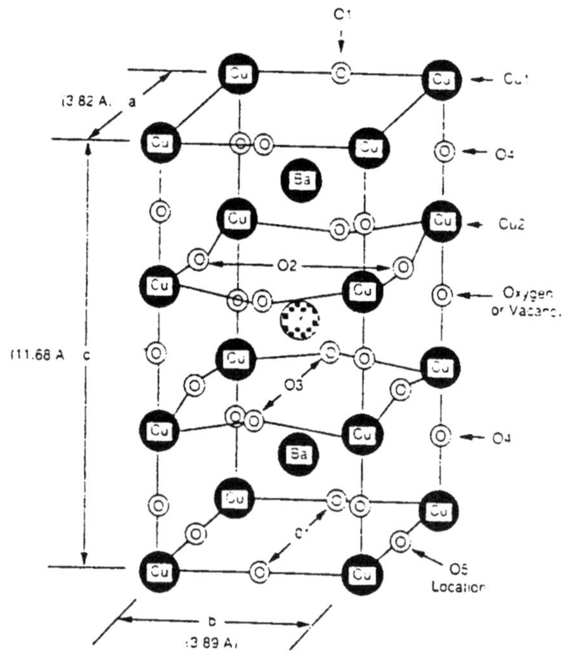

FIGURE 3.4. The $YBa_2Cu_3O_{7-\delta}$ structure showing the five coordinated Cu(2)-O planes and the four coordinated Cu(1)-O chains.

imum T_c approaching 93 K. Note that the lattice parameters of the above-mentioned orthorhombic cell are a = 0.388 nm, b = 0.382 nm and c = 1.168 nm. Although oxygen content is reduced, the orthorhombicity decreases, transforming, therefore, the initial crystal structure to a tetragonal one ($\delta <$ 0.5, $O_{6+6.5}$) and destroying the superconducting properties. Appropriate annealing of the processed YBCO ceramics at low temperatures (450 ÷ 500°C) and in flowing oxygen may reverse the above phase transformation and regenerate the superconductivity of the component.

3.2.3 Bi-Sr-Ca-Cu-O and Tl-Ba-Ca-Cu-O Cuprate Systems

Regarding crystal structure, more complex cuprate superconductors have been found in the discovered Bi-Sr-Ca-Cu-O [7] and Tl-Ba-Ca-Cu-O systems [8]. These structures are generated from the construction of perovskites in combination with rocksalt (NaCl-type) units. The oxides of the above-mentioned systems are members of the homologous series with a generalized formula $A_2B_2Ca_{n-1}Cu_nO_{2n+4}$ (A = Bi, Tl, B = Sr, Ba, n = 1,2,3), and their unit cell is constructed by the sequence of the planes as follows: $(ABO_2)_2$ −

$Ca_{n-1} - (CuO_2)_n$. The intercalation of Ca-layers takes place between the CuO_2 planes. The members of the corresponding homologous series may be characterized briefly by their n value as: $-n = 1$, 2201 phase, $n = 2$, 2212 phase and $n = 3$, 2223 phase.

When $n = 1$, the corresponding 2201 phases for the two systems, namely $Bi_2Sr_2CuO_6$ ($T_c = 10$ K, $a = 0.536$ nm, $b = 0.537$ nm, $c = 2.462$ nm) and $Tl_2Ba_2CuO_6$ ($T_c = 90$ K, $a = 0.546$ nm, $b = 0.547$ nm, $c = 2.323$ nm) are formed by the sequence of stacking $BiSrO_2$-CuO_2-$BiSrO_2$ and $TlBaO_2$-CuO_2-$TlBaO_2$, respectively, see Figure 3.5(a). The structure of the Bi-2201 and Tl-2201 phases are illustrated in Figures 3.6(a) and 3.7(a), respectively.

When $n = 2$, the corresponding 2212 phases for the two systems, namely $Bi_2Sr_2CaCu_2O_8$ ($T_c = 85$ K, $a = 0.540$ nm, $b = 0.542$ nm, $c = 3.093$ nm) and $Tl_2Ba_2CaCu_2O_8$ ($T_c = 115$ K, $a = 0.385$ nm, $c = 2.931$ nm) are formed by the sequence of stacking $BiSrO_2$-$Ca(CuO_2)_2$-$BiSrO_2$ and $TlBaO_2$-$Ca(CuO_2)_2$-$TlBaO_2$, respectively, see Figure 3.5(b). The structure of the Bi-2212 and Tl-2212 phases is illustrated in Figures 3.6(b) and 3.7(b), respectively.

When $n = 3$, the corresponding 2223 phases for the two systems, namely $Bi_2Sr_2Ca_2Cu_3O_{10}$ ($T_c = 110$ K, $a = 0.539$ nm, $b = 0.540$ nm, $c = 3.70$ nm) and $Tl_2Ba_2Ca_2Cu_3O_{10}$ ($T_c = 125$ K, $a = 0.385$ nm, $c = 3.59$ nm) are formed by the sequence of stacking $BiSrO_2$-$Ca_2(CuO_2)_3$-$BiSrO_2$ and $TlBaO_2$-$Ca_2(CuO_2)_3$-$TlBaO_2$, respectively, see Figure 3.5(c). The structure of the Bi-2223 and Tl-2223 phases are illustrated in Figures 3.6(c) and 3.7(c), respectively.

3.2.4 Hg-Ba-Ca-Cu-O Cuprate System

The mercury (Hg)-based cuprate superconductors were reported to have remarkably high Meissner signal onset temperatures [9]. The highest ever re-

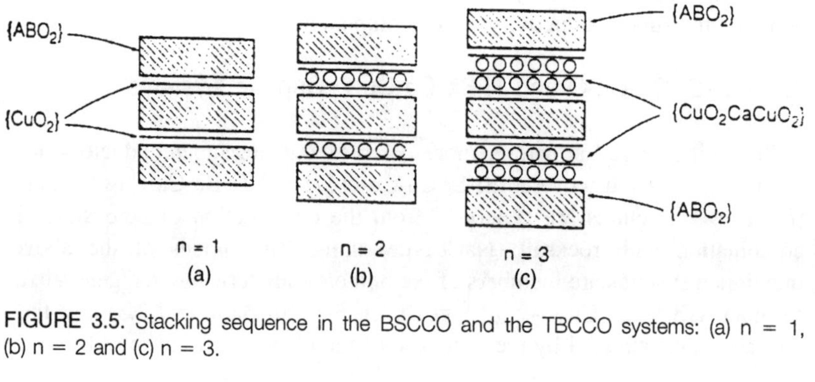

FIGURE 3.5. Stacking sequence in the BSCCO and the TBCCO systems: (a) n = 1, (b) n = 2 and (c) n = 3.

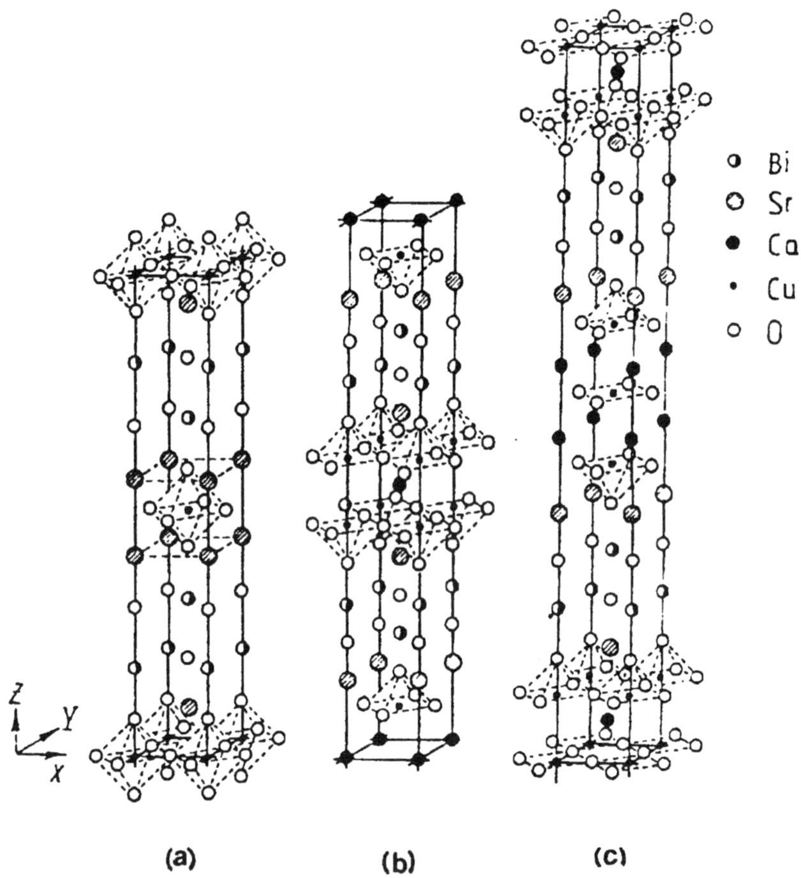

FIGURE 3.6. Unit cells of the members of the BSCCO system: (a) 2201 phase, (b) 2212 phase and (c) 2223 phase.

ported T_c was recorded up to nearly 134 K for the Hg-Ba-Ca-Cu-O system, and it is the highest one for the copper oxide superconducting families. The homologous series of the Hg-Ba-Ca-Cu-O system are described by the formula $HgBa_2Ca_{n-1}Cu_nO_{2n+2+\delta}$ (Hg-12(n − 1)n phases, n = 1,2,3).

The crystal structure of the Hg-12(n − 1)n phases is a sequence of layers of HgO_δ, 2BaO, (n − 1)Ca and $n$$CuO_2$, shown in Figure 3.8. The unit cells of the Hg-1201, Hg-1212 and Hg-1223 correspond to the tetragonal lattice form ($a = b \neq c$, $A = B = C = 90°$) with a = 0.385 nm and c = 0.95 nm, 1.26 nm and 1.57 nm, as observed from HRTEM and electron diffraction patterns [9]. The c-axis length increases with a 0.31-nm step as the number of Ca and CuO_2 planes increases.

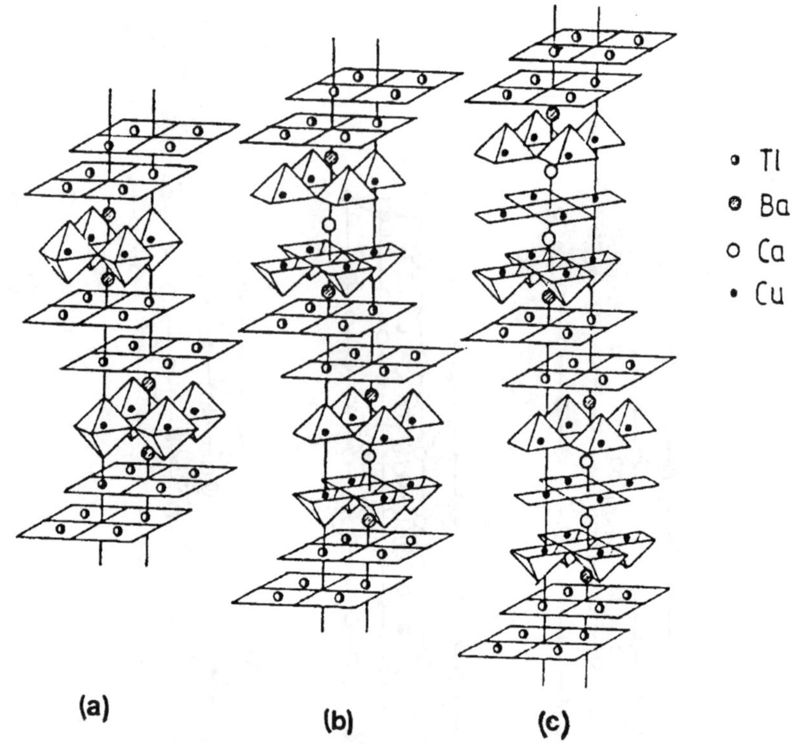

FIGURE 3.7. Unit cells of the TBCCO system: (a) Tl-12(n − 1)n and (b) Tl-22(n − 1)n, n = 1,2,3.

The superconductivity phenomena in the Hg-Ba-Ca-Cu-O system are generated by the charge carriers that exist whenever $\delta \neq 0$. For instance, the superconductivity mechanism in the Hg-1201 phase is driven by the excess oxygen atoms ($\delta > 0$), producing, therefore, holes into the CuO_2 layers, as suggested by Putilin et al. [10]. The critical temperatures of the Hg-1201 phase approaches 95 K, whereas for the Hg-1212 and Hg-1223 phases the corresponding critical temperatures are about 120 and 134 K. The Hg-1201 single-phase ceramic can be produced, while its superconducting properties can be improved by using postannealing in argon (Ar) atmosphere at 700 ÷ 750°C. The Hg-1212 and Hg-1223 phases are found as a superconducting mixture whatever initial stoichiometry is used. The 130 K class superconductivity in the Hg-1212/Hg-1223 phases can be established by additional heat treatment in flowing oxygen at the same temperature range.

The new high-T_c cuprates inaugurate a new era in superconductivity, the so-called *high-temperature* or *high-T_c superconductivity*. The anisotropic structure of these ceramics, viewed as stackings of CuO_2 sheets and separated by layers of al-

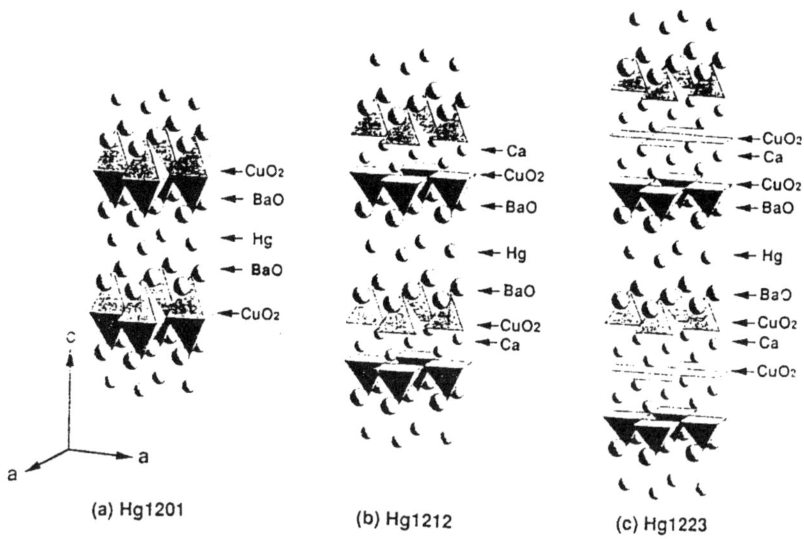

FIGURE 3.8. Crystal structure of (a) Hg-1201 phase ($HgBa_2CuO_4$), (b) Hg-1212 phase ($HgBa_2CaCu_2O_6$) and (c) Hg-1223 phase ($HgBa_2Ca_2Cu_3O_8$) (from Reference [9]).

kaline earth or rare earth oxides, provides also an anisotropy in their superconducting properties; current transport is easier through the *ab* planes rather than the *c* axes. The intermediate oxide layers act as *charge reservoirs,* changing, therefore, the spacing between the CuO_2 planes and, most important controlling the hole concentration within these planes. Note, however, that the theories and the mechanisms of this high-temperature superconductivity are still in progress.

3.3 CHEMICAL SYNTHESIS METHODS

The inorganic chemical processing of high-T_c superconductors is the first step in the fabrication procedure, see Figure 1.4. It is also important because the quality (crystal structure, microstructure and superconducting properties) of the produced material may dramatically affect the soundness and the workability of the final superconducting component. An extensive review presenting the most important chemical synthesis routes was made by Rao et al., see Reference [11].

3.3.1 Solid-State Reaction (Ceramic) Method

The solid-state process is the most common technique for the production of inorganic solids at elevated temperatures, starting from the initial mixture of the raw materials that exist at the solid state (ceramic method) [12]. If one of the constituents is volatile or sensitive to the atmosphere, the chemical process is performed in sealed evacuated containers. The starting materials are metal oxides or carbonates or other salts, which are mixed, homogenized and heated at a given temperature for a relatively long time to achieve maximum reaction conversion.

The ceramic method is the most usual technique for the preparation of superconducting oxides. The main disadvantages of this method are:

- The initial composition is nonuniform throughout the whole mixture of the raw materials.
- It is rather difficult to control the conversion of the chemical reaction to provide the necessary heating duration. This is due to a lot of samples, prepared by this technique, contain a mixture of the reactants and products, whose separation is almost impossible.
- Although no-melt is formed during the chemical reaction, the entire solid-state reaction takes place by the advancement of a reaction front, beginning at the interface boundaries between the particles and progressing by diffusion of the constituents through the product phase. While this reaction is further extended, the diffusion paths become longer and the overall process rate is slower. The reaction can be accelerated by using intermediate grindings between heating cycles.
- In multiphase oxide systems, such as in the case of Bi-Sr-Ca-Cu-O and Hg-Ba-Ca-Cu-O, it is highly difficult to obtain a homogeneous or single-phase product.

A variation of the above-mentioned technique consists of heating of nitrates instead of oxides and carbonates. This technique is based on making metal nitrates, which compose the superconducting compound, by forming a solution of the initial oxides/carbonates in concentrated HNO_3. By evaporating the solvent and drying the solid residue, the corresponding mixture of metal nitrates is obtained. With use of nitrates instead of carbonates, the formation carbon film, covering the superconducting grains and leading to very low critical current densities, J_c, is inhibited.

Y-Ba-Cu-O

Single-phase superconducting compounds, with orthorhombic crystal structure corresponding to $YBa_2Cu_3O_{7-\delta}$, can be obtained by using a stoichiometric mixture of Y_2O_3, CuO and $BaCO_3$ (1:2:3 mol ratio). In the method used for the preparation of the $YBa_2Cu_3O_7$ compound, the initial components are ground thoroughly and heated initially at 950°C in the powder form for 24 h. After the calcination step, the powder is reground, homogenized and sintered at the same temperature for another 24-h period. Finally, annealing in flowing oxygen is performed at about 500°C for 24 h, to obtain the orthorhombic

123 phase, with a superconducting transition at 90 K. Impurity side products, known also as secondary phases, are obtained during the preparation process. The most usual secondary phases are the $BaCuO_2$, Y_2BaCuO_5 (211 or "green" phase) and $Y_2Cu_2O_5$, as shown in the relevant ternary phase diagram of Y_2O_3-BaO-CuO, presented in Figure 3.9; see also Reference [13].

Bi-Sr-Ca-Cu-O

Although the ceramic method is widely used for the synthesis of Bi-based superconducting cuprates, it is generally difficult, or even impossible, to obtain only single-phase material. The 2201 phase ($Bi_2Sr_2CuO_6$) appears to be more stable at 810°C whereas the low-T_c 2212 phase ($Bi_2Sr_2CaCu_2O_8$) becomes stable around 840°C. The high-T_c 2223 phase ($Bi_2Sr_2Ca_2Cu_3O_{10}$) can be obtained close to the melting temperature, i.e., 850°C, after a heat treatment for very long time periods, varying from several days to a few weeks; the 2212 phase seems to be more stable compared with the other members of the Bi-Sr-Ca-Cu-O family. It may be noted, that one of the dominant problems of the synthetic process is the volatile character of the Bi_2O_3 compound, which starts to melt around 830°C. Heating above this temperature, the bismuth oxide is evaporated and, therefore, stoichiometric variations lead to microinhomogeneities and to the presence of unreacted oxides. Because this system contains many cations, partial reactions that occurred between pairs of metal oxides may lead to impurity phases and to final product contamination.

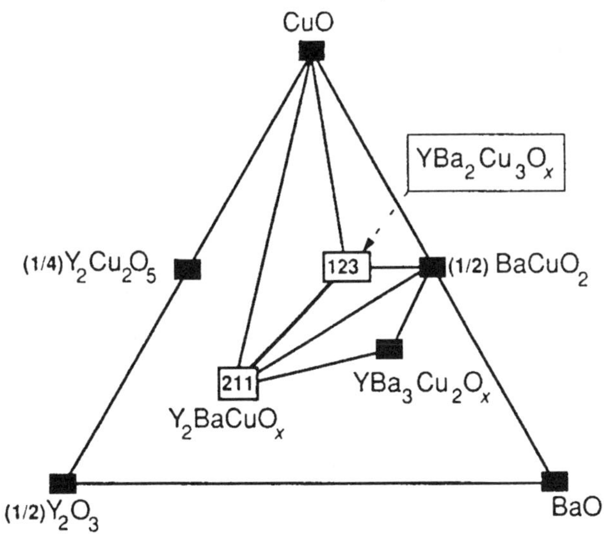

FIGURE 3.9. Ternary phase diagram of the Y_2O_3-BaO-CuO (from Reference [13]).

According to the suggested solid-state reaction technique, called *matrix reaction method,* the number of the reactants is reduced by reacting the oxide matrix, made from the $CaCO_3$, CuO and $SrCO_3$ mixture, with Bi_2O_3. The heating temperature range lies between 810 and 850°C for a minimum time period of 48 h. Successive quenching in air to the room temperature yields to the formation of single 2212 or 2223 phases with T_c 85 and 110 K, respectively. Partial melting of the mixture for a short time period, up to 5–10 min, may lead to the rapid formation of the 2212 and 2223 phases.

Lead addition, in the form of PbO, favors the kinetics of the formation of the 2212 and 2223 phases. Bismuth (Bi) is partially substituted by lead (Pb), forming a superconducting solid solution, whereas the formation of the so-called *liquid phase,* namely the Ca_2PbO_4 with a melting point of about 800°C, is a catalyzing substrate for the growth of the 2223 crystals. By using Bi/Pb ratio equal to 1.6/0.4, an almost single 2223 phase material may be obtained by heat-treating it at nearly 870°C for 5 days using the matrix reaction technique.

Tl-Ba-Ca-Cu-O

Safety precautions may be considered, when handling thallium (Tl)-based materials, because of the increased volatility and toxicity of Tl_2O_3. Synthesis of Tl cuprates, mainly of the $Tl_2Ba_2Ca_{n-1}Cu_nO_{2n+4}$ homologous series, has, therefore, been obtained in closed tubes made usually from gold or nickel alloy or, alternatively, in a sealed crucible or ampule. The reaction takes place between the mixture of the $CaCO_3$, CuO, $BaCO_3$ and Tl_2O_3 (matrix reaction) compounds, to obtain the 2223 phase as a major constituent, besides the 2212 phase. The heating temperature range is 880–930°C and the heat treatment duration ranges from 20 min to 6 h.

Hg-Ba-Ca-Cu-O

The superconducting compounds of the homologous series of $HgBa_2Ca_{n-1}Cu_nO_{2n+2+\delta}$ can be prepared by the solid-state reaction technique in sealed tubes, because of the highly volatile HgO compound. The mixture of the raw materials used is composed of oxides HgO, BaO, CaO and CuO, possessing the desired stoichiometry. The powder mixture is homogenized in a dry container, to prevent CO_2 and water absorption. The homogenized mixture is pelletized and sintered in a quartz tube and evacuated at a pressure of about $10^{-}4$ Torr. The heating temperature ranges from 600 to 800°C and the heat treatment duration from 20 min to 12 h. The production of Hg-1201 monophase ceramics, with T_c = 95 K, is possible, while the separation between the Hg-1212 and Hg-1223 phases is very difficult. Postprocessing annealing in flowing oxygen, at low temperatures of about 300°C for long time periods, 6–72 h, enhances the 130 K class of superconductivity.

3.3.2 Coprecipitation and Precursor Techniques

Coprecipitation is the physicochemical process that deals with the separation of a solid containing various ions from a solution. A homogeneous coprecipitation usually results in the formation of nanometer-scale precipitates of quasi-crystalline or even amorphous form. Coprecipitation of species, possessing strictly defined chemical composition and stoichiometry, can be achieved under the following conditions:

- The precipitating agent is a multivalent organic compound that can coordinate more than one metal ion, whereas the precipitation rate is rapid.
- Careful adjustment of the pH in the solution must be made.
- The solid precipitating out of the solution must be insoluble in matrix liquid.

The anions preferred for coprecipitation of oxides are carbonates, oxalates and citrates. The formed precipitates are very well mixed in the atomic scale, and the resulted ceramic, after the appropriate heat treatment, is characterized by a high bulk density and a very low grain size. After the precipitation, the particles must be heat-treated (firing process) at a suitable temperature, to obtain the desired high-T_c cuprate.

Y-Ba-Cu-O

By using various organic reagents, the orthorhombic 123 superconducting phase of the Y-Ba-Cu-O compound, can be obtained by the coprecipitation technique. The Y-Ba-Cu-O group is the superconducting system that is mostly studied.

The three most familiar basic coprecipitation synthesis routes, as far as the precursors used, are: the oxalate precipitation, the hyponitrite precipitation and the hydroxycarbonate precipitation.

In the case of *oxalate coprecipitation,* an oxalic acid solution is added to an aqueous solution of nitrates of Y, Ba and Cu; the pH of the solution is adjusted to 7.5. The green slurry formed, after precipitation, is filtered and dried. The oxalate precipitates, contained in the slurry, can be converted to $YBa_2Cu_3O_{7-\delta}$ after heat treatment at 780°C in air for 5 days and subsequent postannealing in flowing oxygen at 450°C. By using this certain oxalate coprecipitation technique, stoichiometric variations in the final product may be raised, because of the moderate solubility of the barium oxalate and the formation of $BaCO_3$ during the calcination procedure.

Another technique ensuring the avoidance of $BaCO_3$ formation is the use of the *hyponitrite precursor.* The hyponitrite precursor is formed by adding a solution of sodium hyponitrite ($Na_2N_2O_2$) in a solution of Y, Ba and Cu nitrates. The hyponitrite precipitate is converted to 123 phase by heat treatment at 700°C followed by oxygenation at 400°C. The contamination of the slurry with alkali ions is possible.

The $YBa_2Cu_3O_{7-\delta}$ compound can be obtained by the *hydroxycarbonate route,* by using either KOH/K_2CO_3 or $NaOH/Na_2CO_3$ for the precipitation of Cu hydroxide and the Y/Ba carbonates at pH, varying between 7–8 and 13, respectively; the possibility of the precipitate contamination by alkali ions cannot be excluded.

Bi-Sr-Ca-Cu-O

Few studies have dealt with coprecipitation synthesis, because of the rather complex chemistry of the BSCCO system and the decomposition of bismuth nitrate in cold water.

Monophasic 2223 compound (T_c = 110 K) can be obtained by using the oxalate precipitation technique. Addition of oxalic acid to metal nitrate solution causes successful precipitation at pH equal to 6.7. Careful adjustment of pH is controlled by the NH_4OH solution. The $Bi_{1.6}Pb_{0.4}Sr_2Ca_2Cu_3O_x$ compound can be produced by sintering the precipitates at about 860°C in air for 72 h.

Tl-Ba-Ca-Cu-O

The oxalate precipitation is also suggested here for the production of the high-T_c member 2223 phase (T_c = 125 K). By using this technique a stoichiometric mixture of Tl acetate, $BaCO_3$, $CaCO_3$ and Cu acetate is dissolved in water containing glacial acetic acid. The solution is then added to oxalic acid (excess) under stirring. The oxalate precipitation after filtering, drying and pelletizing is heat-treated in a closed tube in oxygen atmosphere at 900°C for several minutes.

3.3.3 Sol-Gel Method

The *sol-gel process* is an advantageous technique for the production of superconducting powders and films at low sintering temperatures and chemically homogeneous by atomic mixing. The main difference between *sol* and *gel* is that the latter had a tendency to retain its characteristic shape, whereas the former (sol = colloidal dispersion) has the shape of the container. The sol represents the colloidal suspension, whereas the gel represents a colloidal or polymeric solid, which also contains a liquid component homogeneously dispersed in it. In the sol-gel process, a sol containing all the reacting oxides and hydroxides is converted to a gel by solvent removal. The gel is then fired to an appropriate temperature to obtain the product.

Two different variations of the sol-gel process are used for the production of copper oxide superconductors through:

- molecular precursors such as metal alcoxides in organic medium
- ionic precursors in aqueous medium, e.g., citrate sol-gel processing

It may be noted that the chemical composition and the microstructure of the final products are affected by the main parameters of the sol-gel process, i.e., the solvent, the pH and the firing temperature.

Y-Ba-Cu-O

Improved 123 YBCO compound can be sol-gel synthesized by *n*-butoxides of Y, Ba and Cu in butanol solvent. The organometallic precursor, after drying, is sintered at 700°C and then oxygenated at 400°C, to achieve the orthorhombic 123 phase.

The superconducting YBCO compound may be also produced by the citrate sol-gel route, by adding an equivalent amount of citric acid to the metal ions. The pH is carefully controlled by addition of NH_4OH or ethylenediamine. After the evaporation of the solvent, a relatively dark blue gel is obtained. After the decomposition of the gel, the powder is sintered at 900°C in flowing oxygen, yielding to orthorhombic $YBa_2Cu_3O_{7-\delta}$ with submicrometer grain size of about 0.5 μm.

$BaCO_3$ formation during calcination and alkali metal ions contamination of the final material is still problematic. The sol-gel process is of great technological importance, due to the application of this method, not only for powder synthesis but also for the manufacture of useful superconducting components, such as thin/thick films, fibers, etc.

Bi-Sr-Ca-Cu-O

The sol-gel process can be successfully applied for the synthesis of highly stoichiometric and, therefore, of almost monophasic Pb-doped 2223 BSCCO ceramics.

The working solution is composed of the initial metal (Bi, Pb, Sr, Ca, Cu) nitrates aqueous solution and equivalent amount of EDTA (ethylene-diamine-tetra-acetic acid). The pH is carefully adjusted to 8 by adding NH_4OH. The reactions that take place yield to certain organometallic complex precursors [14]:

$$H_4EDTA + 4NH_3 \rightarrow EDTA^{4-} + 4NH_4^+ \quad (3.1)$$

$$EDTA^{4-} + Bi(NO_3)_3 \rightarrow [Bi(EDTA)]^- + 3\ NO_3^- \quad (3.2)$$

$$EDTA^{4-} + M(NO_3)_2 \rightarrow [M(EDTA)]^{2-} + 2\ NO_3^- \quad (3.3)$$

$$(M: Pb^{2+}, Sr^{2+}, Ca^{2+}, Cu^{2+})$$

$$Cu(NO_3)_2 + 4NH_3 \rightarrow [Cu(NH_3)_4]^{2+} + 2NO_3^- \quad (3.4)$$

After the drying process, the gels are subjected to thermal decomposition by firing in oxygen at 520°C for 10 h. The resulted solid precursors are calcined up to 800°C in air for 12 h. The produced powder is sintered at a temperature range of 840–860°C in air for a time period varying between 50 and 300 h. The 2223 phase is the major constituent of the fabricated ceramic powder over a percentage of 92% w/w with a T_c equal to 106 K.

3.3.4 Emulsion Technique

YBCO superconducting powders, possessing spherical monosized morphology of nanometer dimensions (100–1000 nm), can be synthesized by the *emulsion technique* [15]. According to this technique, an organic fluid containing a surfactant is added to the aqueous solution of the ceramic precursors. The organic fluid causes the formation of homogeneously dispersed water droplets of uniform size in the organic phase (emulsion formation). Each one of the water droplets contains the same amount of the ceramic precursors. The aqueous and the oil phases are mixed in the proper ratios in a high-shear rate mixer. To complete the emulsion processing, the premixed solution is treated through a two-stage homogenizer, operating at 34 MPa, and, therefore, the uniformity and the nanometer scale size of the particles can be achieved.

The separation of the solid precursors from the fluid can be accomplished by using vacuum distillation techniques. The resulted slurry of the solid particles is subjected to a firing process at 500°C to remove the organics. Calcination at 900°C for 5 h in oxygen atmosphere is necessary for the formation of the YBCO superconducting phase. Traces of the $BaCO_3$ impurity phase are detected in the final product. The presence of certain dopant compounds, namely B_2O_3, reduces the calcination temperature from 900 to 550°C.

3.3.5 Self-Propagating High-Temperature Synthesis (SHS)

Self-propagating high-temperature synthesis (SHS), known also as *combustion synthesis,* was developed for the production of inorganic materials of

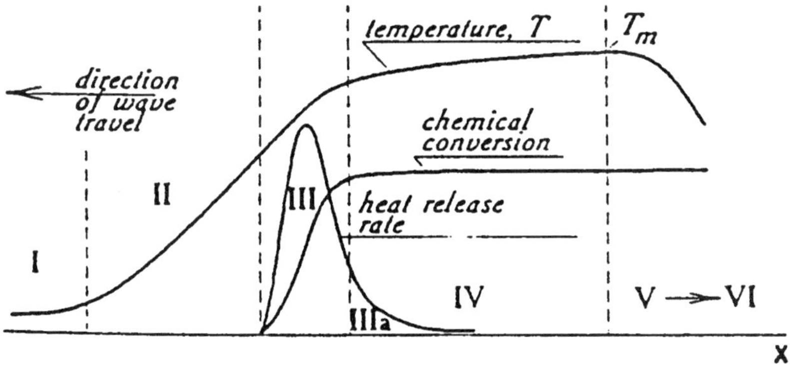

FIGURE 3.10. Wave structure in the SHS process. I: reactants; II: preflame zone; III: zone of main heat release; IIIa: zone of afterburning; IV: chemical conversion zone, V: cooling of products; VI: final SHS products (from Reference [16]).

great technological importance [16]. It is also a simple and rapid technique of preparation of high-T_c superconducting oxides, based on the thermal energy released by many highly exothermic solid-solid and gas-solid noncatalytic reactions. This energy release rate may trigger the reaction propagation under very high rates. The above-mentioned process occurs in a very narrow zone that separates the reactants from the products.

By controlling the combustion velocity and temperature, the chemical composition and structure of the final products can be easily adjusted, leading very often to high-quality chemical synthesis. The combustion wave of SHS process has, generally, a very complex structure, see Figure 3.10. The combustion zone that separates the preflame zone and the zone of chemical conversion is an important element of this structure. The heat release, which influences the front propagation, occurs in the chemical conversion zone that is adjacent to the combustion front. In the preflame zone, intense heat transfer takes place, but chemical conversion is not still occurring. The edge component of the chemical conversion, namely the afterburning zone, helps the fast propagation of the transformation processes to the reactants.

The overall SHS mechanism of the YBCO 123 phase formation is as follows:

(1) Warm and combustion zones (maximum temperature: 830°C)

$$2Cu + O_2 \rightarrow Cu_2O + 1/2O_2 \rightarrow 2\ CuO \tag{3.5}$$

$$BaO_2 \longleftrightarrow BaO + 1/2O_2 \tag{3.6}$$

$$BaO + BaO_2 \rightarrow L \tag{3.7}$$

$$L + 2Cu_2O_2 \rightarrow 2BaCu_2O_2 \quad (3.8)$$

$$L + 2CuO \rightarrow 2BaCuO_2 \quad (3.9)$$

(2) Post-process zones (maximum temperature: 950°C)

$$2BaCu_2O_2 + CuO + 1/2Y_2O_3 \rightarrow YBa_2Cu_3O_{6+x} + Cu_2O \quad (3.10)$$

$$BaCuO_2 + CuO \rightarrow BaCu_2O_2 + 1/2O_2 \quad (3.11)$$

$$2BaCuO_2 + 1/2Y_2O_3 + CuO \rightarrow YBa_2Cu_3O_{6+x} \quad (3.12)$$

(3) Cooling zones

$$YBa_2Cu_3O_{6+x} + yO_2 \rightarrow YBa_2Cu_3O_{7-\delta} \quad (3.13)$$

3.3.6 Other Techniques

Minor synthetic techniques, such as freeze drying, spray drying and electrochemical methods, are used for the production of cuprate superconductors. In spray drying, a solution of metal nitrates is sprayed in the form of fine droplets into a hot chamber. Drying for solvent removal, followed by appropriate heat treatment, may lead to the formation of the 123-YBCO compound and the 2223 (major) phase material. Similar to spray drying is the aerosol process for the manufacture of ultrafine superconducting powders [17].

3.4 EFFECT OF DOPING ON THE STRUCTURE AND PROPERTIES OF HTS

The transport properties of ceramic superconductors are strongly affected by the microstructure of these materials and, mainly, by the nano- and microscale defect structure, i.e., microcracks, crystallographic misorientation, dislocations/lattice disorder and grain boundaries. The latter is a very important parameter resulting in the reduction of J_c in granular superconducting materials, due to Josephson weak-links formation between the grains. The existence of doping elements in substitution lattice sites lead to the formation of a superconducting solid solution with altered, in many cases improved, superconducting properties. The influence of dopants on the copper oxide HTS microstructure and superconducting properties has gained a great deal of attention. In the following sections, the effects of the various important doping elements on the structure and superconducting properties are examined in the cases of the most widely studied superconducting systems, mainly the Y-Ba-Cu-O and Bi-Sr-Ca-Cu-O ceramics.

3.4.1 Y-Ba-Cu-O Cuprate System

DOPING WITH ALKALI ELEMENTS

Regarding the oxygen content and the heat treatment, the sensitivity of the superconducting properties of the compound $YBa_2Cu_3O_y$ can be reduced by substituting barium with alkali cations (Li, K, Na). The presence of extrinsic dopants results in fixing the oxidation state of copper, reducing, therefore, the sensitivity to oxygen activity and temperature. The replacement of Ba^{2+} by K^+ ions and/or Na^+ ions, resulting in an oxygen content (y) higher than 6.8, favors the increase of the T_c and the formation/stabilization of the 123 orthorhombic superconducting phase in a less oxidizing atmosphere [18].

The addition of KF in the raw oxide materials mixture aims to the partial replacement of Ba in the crystal lattice of the 123 compound. Doping the Y-Ba-Cu-O ceramic with alkali cations (Li, Na, K) favors the stabilization of the 123 orthorhombic phase ($YBa_2Cu_2O_{7-\delta}$) at lower sintering temperatures and durations, leading to higher transition temperatures up to 100 K. This can be explained by the comparative X-ray photoelectron spectroscopy (XPS) study of nondoped and K-doped samples [19]; in the latter case, a greater amount of trivalent copper ions was found according to the following reaction:

$$Cu^{2+} + Ba^{2+} \rightarrow K^+ + Cu^{3+} \tag{3.14}$$

Potassium atoms form easier a superconducting solid solution by partial barium substitution $YBa_{2-x}K_xCu_3O_{7-\delta}$ (x = 0–0.1), than other alkali atoms (K > Na > Li). This is due to the position of the above alkali elements in the periodic table, relative to barium, explaining, therefore, the potassium greatest compatibility with barium, which has the nearest atomic radius (0.227 nm), compared to barium (0.217 nm), and the same coordination number (CN = 6).

Potassium addition affects strongly the coupling phenomena among the grains, resulting in improving the quality of the intergranular links and, therefore, in increasing the critical current density. The optimum K content (x) seems to be equal to 0.40; above this critical content, a sufficient grain size reduction takes place, leading to subsequent intergranular decoupling and deteriorating the superconducting properties [20].

The formation of mixed K-Cu oxide phases, such as K_3CuO_2, confirmed for high K contents by X-ray diffraction, strongly affects the grain size and the coupling quality [20].

Evidence of the partial substitution of Ba by K atoms can be shown by taking into account the Raman spectra of the doped YBCO ($x = 0.05$), see Figure 3.11. These spectra correspond to the fully oxygenated ceramic, with oxygen content up to 6.93. The peaks, observed at a wavenumber near 117 cm^{-1}, correspond to vibrations of potassium atoms that partially occupy the Ba sites in the lattice. The relatively high T_c of these samples may be attributed to the increase of the carrier concentration in the Cu-O planes.

On the other hand, by adding KF in the oxide mixture, oxygen ions can be partially replaced by fluoride, which is a more electronegative atom, reducing the sensitivity of the YBCO compounds to oxygen content. These two substitutions, i.e., Ba with K and O by F, result in an increase of the T_c by 6–10 K and of the J_c at 77 K [18].

CADMIUM DOPING

Alloying of Y-Ba-Cu-O ceramics ($YBa_2Cu_3Cd_xO_y$), in a content higher than the solubility limit ($x > 0.2$), results in the deterioration of the Josephson link network because of the presence of secondary Cd-rich phase at the grain boundaries. This can, therefore, result in the reduction of superconducting properties, i.e., the critical current density and the diamagnetic shielding at low temperatures. On the other hand, a Cd content below the solubility limit does not affect the transition temperature and is expected to enhance the stability of the superconducting phase, see Reference [21].

NIOBIUM DOPING

The single phase c-$YBa_2Cu_{3-x}Nb_xO_y$ ($x = 0.05$–0.20) with cubic symmetry can be synthesized by thermolysis of an ammonium nitrate melt containing quantities of Y, Ba, Cu and Nb elements [22]. Nb atoms substitute Cu in the lattice and lead to the formation of a cubic phase, which is very similar to 143 phase in the Y-Ba-Cu-O system ($YBa_4Cu_3O_y$), containing 50% Ba vacancies. The transformation of the cubic phase to the orthorhombic one results in the formation of a superconducting material with very high values of J_c.

TUNGSTEN DOPING

Tungsten (W) addition in the Y-Ba-Cu-O system creates a substitional solid solution of the type $YBa_2Cu_{3-x}W_xO_{9-\delta}$. For W content $x < 0.2$, a simple perovskite lattice is detected by XRD, whereas for $x > 0.2$ a face-centered perovskite unit cell is derived. Even small W contents lead to the destruction of the Ba-Y-Ba ordering in the 123 phase and to the elimination of the superconductivity [23].

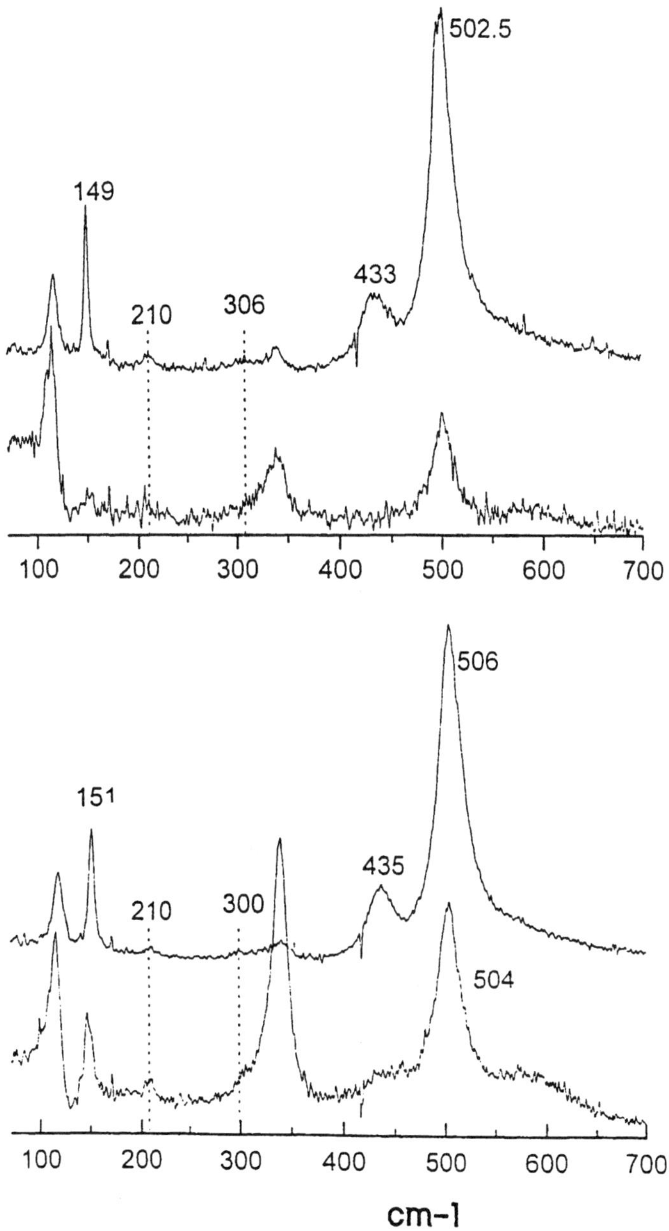

FIGURE 3.11. Raman spectra of two different crystallites of a K-doped YBCO sample.

3.4.2 Bi-Sr-Ca-Cu-O Cuprate System

LEAD DOPING

Lead (Pb) is the most important doping element that influences the microstructure, phase composition and the related superconducting properties of the Bi-Sr-Ca-Cu-O system. The presence of Pb in the initial mixture, usually as PbO, favors the reaction kinetics of the 110 K 2223 phase via the formation of the "liquid" Ca_2PbO_4 phase. Note that various possible 2223 phase formation mechanisms have been proposed, namely the *intercalation,* the *disproportionation* and the *dissolution-precipitation* mechanism.

Lead addition results in the creation of a superconducting solid solution $Bi_{2-x}Pb_xSr_2Ca_2Cu_3O_v$, by partial Bi substitution. The optimum Pb content lies between 0.3 and 0.4, according to Reference [24]. The enhancement of the 2223 phase can be derived by comparing the characteristic crystallographic reflection peaks (002) at $2\theta = 4.7°$, taken by the corresponding XRD patterns for various Pb contents, see Figure 3.12. The BSCCO samples were prepared by successive prolong sintering steps.

The Pb concentration also affects the intergrain connections, which constitute a very important factor, dealing with the critical current density and the current carrying capacity of the material. An optimal Pb content, x, around 0.3 is established, which leads to the formation of strong links and coupling between the grains, resulting in sharp diamagnetic transition [24].

DOPING WITH V_B ELEMENTS

The presence of high valency cations V^{5+}, Nb^{5+} and Ta^{5+} of the V_B group in the initial stoichiometry ($Bi_{2-x}M_xSr_2Ca_2Cu_3O_v$, M: V,Nb,Ta) can significantly enhance the formation of the high-T_c 2223 phase. Their role is quite similar to that of Pb for stabilization of the 110 K phase. The optimum doping level, determined from ac-susceptibility and resistivity measurements, is around 0.4 [25].

Alterations in the crystal structure take place by doping with high-valency cations: a reversible transformation undergoes from the annealing process to the quenching process, implying great changes in Bi-O layer:

$$2\,Bi^{3+} \leftrightarrow Bi^{5+} + Bi^{+} \tag{3.15}$$

These elements may also partially substitute, besides Bi, also Cu. This substitution may enhance the stability of the 2223 phase because of the extra oxygen incorporation and the subsequent reinforcement of interlayer bonding. The XRD patterns indicate impurity phase traces, such as $Sr_{0.82}NbO_3$, $Sr_4V_2O_9$ and $Sr_3Ta_2CuO_9$, for Nb-, V- and Ta-doped BSCCO ceramics, respectively.

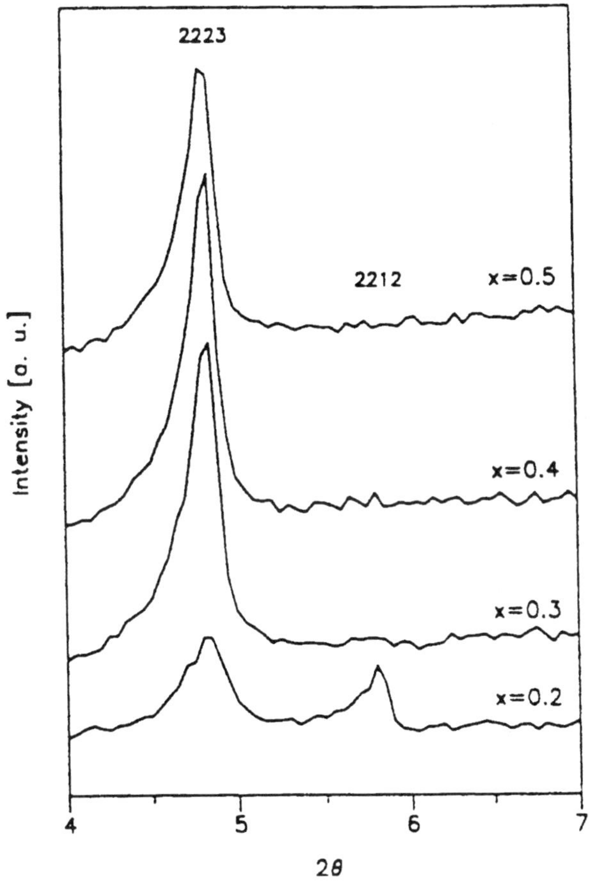

FIGURE 3.12. XRD patterns of $Bi_{2-x}Pb_xSr_2Ca_2Cu_3O_v$ samples: x = 0.2, x = 0.3, x = 0.4 and x = 0.5 (from Reference [24]).

3.5 THERMAL TREATMENT OF HTS POWDERS

3.5.1 General

Thermal (heat) treatment is very often the final stage of fabrication processing of high-temperature superconductors. High-temperature thermal processing reduces the porosity and, therefore, increases the bulk density of the ceramic materials, enhancing their mechanical and physical properties. Careful selection of the appropriate thermal cycle could also improve the crystallinity of some ceramic oxides in amorphous or glass state; see for example melt-quenched ceramics. Oxygen content is also influenced by the heat treatment leading to important modifications in superconducting properties and mainly in T_c.

In the Y-Ba-Cu-O system the oxygen stoichiometry greatly affects superconductivity. The deficiency δ increases by heating the 123 compound above 550°C in air, leading to the transformation of the orthorhombic crystal structure to a tetragonal one. However, absorption of oxygen can be realized by performing heat treatment at temperatures ranging from 400 to 900°C in oxygen atmosphere; the optimum heating temperature lies between 450 and 550°C.

In the Bi-Sr-Ca-Cu-O system the superconductivity is less sensitive to the oxygen stoichiometry compared with the previous YBCO ceramics. Note, however, that for oxygen stoichiometry, the effect of the 2223 phase is quite the opposite than that of the 2212 phase; the increase of oxygen content of the 2223 phase leads to increase in T_c, whereas the opposite is observed for the 2212 phase. This dissimilar behavior of the BSCCO system compared with the YBCO system may be explained by the different hole density effect developed by the more complicated Bi-based structure. The peculiarities observed in the BSCCO compounds for the effect of oxygen content are probably based to the following reasons:

- carrier density changes
- hole distribution between Cu-O and Bi-O planes and
- structural modulation and distortion in Bi-O layers

3.5.2 Chemical Reactions and Transformations

Y-Ba-Cu-O SYSTEM

The main transformation occurred in the solid state is the conversion of the orthorhombic crystal structure to a tetragonal one:

500 < T < 1010°C

$$YBa_2Cu_3O_{7-\delta}\ \text{(ortho-)} \leftrightarrow YBa_2Cu_3O_6\ \text{(tetr.)} + (1-\delta)/2\ O_2 \quad (3.16)$$

Incogruent melting of 123 compound takes place above the peritectic temperature (1010°C) [13,27] :

1010 < T < 1300°C

$$YBa_2Cu_3O_x \rightarrow \text{Liquid} + Y_2BaCuO_5\ \text{(green phase)} \quad (3.17)$$

Higher temperatures lead to further decomposition of the so-called green phase to yttrium oxide:

1300 < T < 1500°C

$$Y_2BaCuO_5 + \text{Liquid} \rightarrow Y_2O_3 \text{ (brown phase)} + \text{Liquid} \quad (3.18)$$

Heat treatment is used for solid-state processing and, therefore, the corresponding reaction is described by Equation (3.16). The chemical equation, which is a reversible process, concerns the oxygenation treatment and the regeneration of the superconducting orthorhombic phase in oxygen atmosphere. All the other chemical reactions take place in melt processing of YBCO ceramics.

Bi-Pb-Sr-Ca-Cu-O SYSTEM

The reaction mechanism scheme is rather complicated for this Bi-based multicomponent system, see References [14], [28] and [29]. To clarify the elementary chemical transformations the following reactions can be written:

770 < T < 800°C (2212 Phase Formation)

$$Bi_2Sr_2CuO_6 + CaO + CuO \rightarrow Bi_2Sr_2CaCu_2O_8 \quad (3.19)$$

T ≈ 810°C (Melting of Ca_2PbO_4)

$$Ca_2PbO_4 \rightarrow \text{Liquid} + CaO \quad (3.20)$$

830 < T < 845°C (Precipitation of 2201 Phase)

$$Bi_2Sr_2CaCu_2O_8 + \text{Liquid} \rightarrow Bi_2Sr_2CuO_6 + \text{Liquid} \quad (3.21)$$

840 < T < 860°C (2223 Phase Formation)

$$Bi_2Sr_2CuO_6 + 2(CaO + CuO)_{\text{Liquid}} \rightarrow Bi_2Sr_2Ca_2Cu_3O_{10} \quad (3.22)$$

The presented reaction scheme, Equations (3.19)–(3.22), constitutes the *dissolution-precipitation mechanism.* Note, however, that according to the *disproportioning mechanism,* the 2223 phase is directly formed from the 2212 phase at the same temperature range:

$$2Bi_2Sr_2Ca_2Cu_3O_8 \rightarrow Bi_2Sr_2Ca_2Cu_3O_{10} + Bi_2Sr_2CuO_6 \quad (3.23)$$

The dissolution-precipitation and the disproportioning mechanisms are the most important reactions dealing with the 2223 phase formation.

3.6 NEW SUPERCONDUCTING COMPOUNDS: AN OUTLOOK

3.6.1 Crystal Structure

The superconducting oxycarbonates present a relatively complicated crystal structure. Their structure is constructed by the combination of two different units, mainly the $Sr_2CuO_2CO_3$ (S2CC) and $A_{1-x}M_xSr_2CuO_{5-\delta}$ (A: Tl, Hg, M: Bi, Cr, Mo, V, Pb) (1201) sections [26].

The first class is simply resulted from the intergrowth of the 1201-type layer with one S_2CC-type layer and can be denoted as [1201][S_2CC]. The first member of this type of materials, isolated by Huvé [30], is the $Tl_{0.5}Pb_{0.5}Sr_4Cu_2O_7CO_3$, represented in Figure 3.13(a). Several other oxycarbonates were isolated under the generalized formula $A_{1-x}M_xSr_4CuO_{7-\delta}CO_3$ (A: Tl, Hg, M: Bi, Cr, Mo, V, Pb).

The two other classes of oxycarbonates are derived from the [1201][S_2CC] family group by applying a periodic crystallographic shearing, transversal with respect to the Cu-O planes, either in the (100) or in the (110) plane of the perovskite; these compounds are, therefore, called *(100)* and *(110) collapsed oxycarbonates,* respectively. The (100) collapsed oxycarbonate corresponds to $TlSr_2Ba_2Cu_2O_7CO_3$, see Figure 3.13(b) and is generated by a crystallographic shear plane, developed every CuO_6 octahedra. In a similar manner, the (110) collapsed oxycarbonate is produced by crystallographic shearing of the [1201][S_2CC] structure every five octahedra, for the compound shown in Figure 3.13(c).

3.6.2 Superconducting Properties

All these collapsed oxycarbonate structures possess superconducting transitions with T_c varying between 60 and 77 K. For the $Hg_{1-x}V_xSr_{4-y}Ba_yCu_2O_7CO_3$-type oxycarbonates the T_c is affected by the vanadium content and by the ratio Sr/Ba. The V content decreases as the Ba content increases, and the T_c decreases as the ratio Sr/Ba decreases. The shearing phenomena lead, in many cases, to the destruction on superconductivity; the rupture between the CuO_2 planes has a poisonous effect on superconducting properties.

The deeper knowledge and understanding of the building of these peculiar structures and the mechanisms of the related superconduction can be very useful to the creation of new and more stable superconducting materials, capable to control and modify their properties.

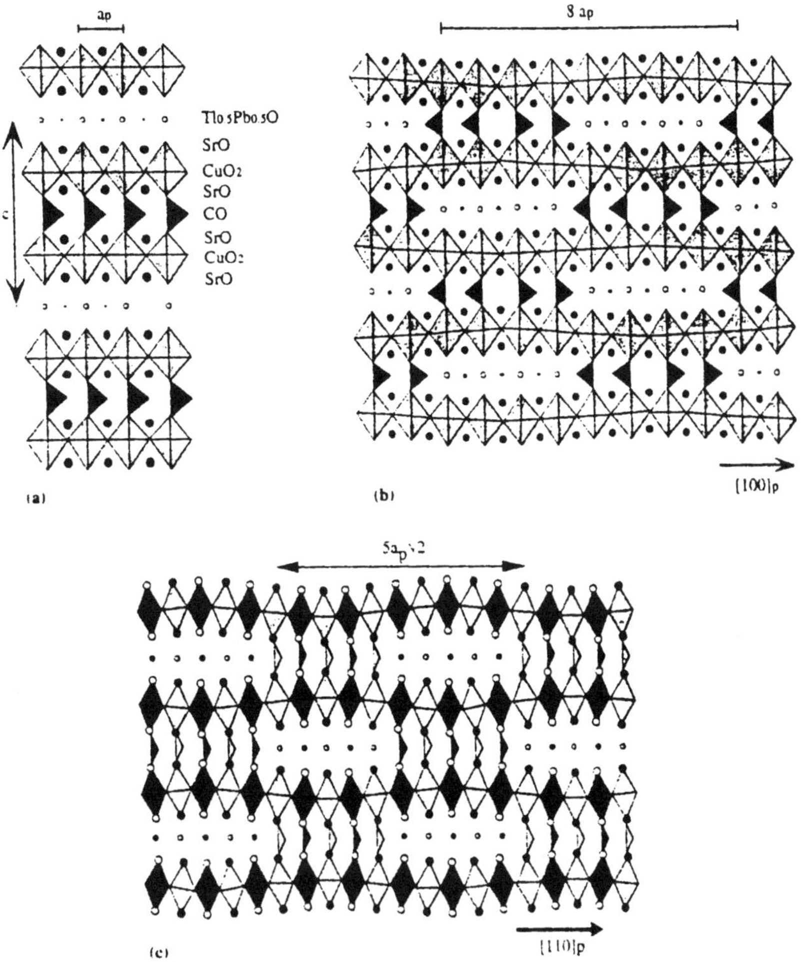

FIGURE 3.13. Idealized structures of the oxycarbonates: (a) $Tl_{0.5}Pb_{0.5}Sr_2Cu_2O_7CO_3$, (b) (100) collapsed $TlSr_2Ba_2Cu_2O_7CO_3$ and (c) (110) collapsed $Tl_{0.8}V_{0.8}Sr_4Cu_2O_7CO_3$ (from Reference [30]).

3.7 REFERENCES

1. Bednorz J. G. and Müller K.A. (1986), "Possible high-T_c superconductivity in the Ba-La-Cu-O system," *Z. Phys.* B64, 189.
2. Smart L. and Moore E. (1993), *Solid State Chemistry,* Chapman & Hall, London.
3. Chiang Y.-M., Birnie D. and Kingery W. D. (1997), *Physical Ceramics: Principles of Ceramic Science and Engineering,* John Wiley & Sons, New York.

4. Raveau B., Michel C. and Hervieu M. (1990), *Studies of High-Temperature Superconductivity,* Nava Science Publishers, New York.

5. Aleksandrov K. S. and Beznosikov B. V. (1997), *Perovskite Crystals,* Hayka Publishers, (in Russian).

6. Wu M. K., Ashburn J. R., Tong C. J., Meng R. L., Gao L., Huang Z. J., Wang Y. Q. and Chu C. W. (1987), "Superconductivity at 93 K in a new mixed phase Y-Ba-Cu-O compound system at ambient pressure," *Phys. Rev. Lett.* 58, 908.

7. Maeda H., Tanaka Y., Fukutomi M. and Asano T. (1988), "A new high-T_c oxide superconductor without a rare earth element," *Jap. J. Appl. Phys. Lett.* 27, 209.

8. Sheng Z. Z., Kiehl W., Bennet J., El Ali A., Marsh D., Mooney G. D., Arammash F., Smith J., Viar D. and Hermann A. M. (1988), "New 120 K Tl-Ba-Ca-Cu-O superconductor," *Appl. Phys. Lett.* 52, 1738.

9. Tokiwa-Yamamoto A., Isawa K., Itoh M., Adachi S. and Yamauchi H. (1993), "Composition, crystal structure and superconducting properties of Hg-Ba-Cu-O and Hg-Ba-Ca-Cu-O superconductors," *Physica C* 216, 250.

10. Putilin S. N., Antipov E. V., Chmaissen O. and Marezio M. (1993), "Superconductivity at 94 K in $HgBa_2CuO_{4+\delta}$," *Nature* 362, 226.

11. Rao C. N. R., Nagarajan R. and Vijayaraghavan R. (1993), "Synthesis of cuprate superconductors," *Supercond. Sci. Technol.* 6, 1.

12. Rao C. N. R. and Gopalakrishnan J. (1989), *New Directions in Solid State Chemistry,* Cambridge University Press, Cambridge.

13. Oka K., Nakane K., Ito M., Saito M. and Unoki H. (1988), "Phase-equilibrium diagram in the ternary system Y_2O_3-BaO-CuO," *Jap. J. Appl. Phys. Lett.* 27, 1065.

14. Przybylski K. (1995), "The chemistry of Bi-based HTSC materials," *Proc. 1st Int. Summer School on High Temperature Superconductivity,* Eger, Hungary, 25.

15. Maher G. H., Hutcins C. E. and Ross S. D. (1996), "Preparation and characterization of ceramic fine powders by the emulsion process," *J. Mater. Proc. Technol.* 56, 200.

16. Merzhanov A. G. (1996), "Combustion processes that synthesize materials," *J. Mater. Proc. Technol.* 56, 222.

17. Tirumala S., Lee D. F., Kroeger D. M. and Salama K. (1997), "Thermomechanical processing and reaction kinetics of Bi-2223 powder-in-tube tapes made from aerosol precursor," *Supercond. Sci. Technol.* 10, 686.

18. Kotsis I. and Enisz M. (1992), "Effect of alkali fluoride additives on the properties of superconductors of the system Y-Ba-Cu-O," *Proc. 8th SIMCER Int. Symp. on Ceramics,* Rimini, Italy.

19. Kotsis I., Enisz M., Oravetz D. and Szalay A. (1995), "Effect of porosity on properties of explosively compacted high-T_c superconductors," *Hung. J. Ind. Chem. Veszprém* 23, 69.

20. Veneva A., Petrov D. K., Dittrich P. and Naughton M. J. (1996), "AC susceptibility and microstructure of alkali doped polycrystalline YBCO HTSC materials," *Physica C* 271, 230.

21. Serfoso G., Kiss L. F., Zsoldos E., Toth J., Sandor S., Papadimitriou L. and Dozsa L. (1994), "On the magnetic properties of cadmium-doped Y-Ba-Cu-O high-temperature superconductors," *J. Mater. Sci. Lett.* 13, 693.

22. Strukova G. K., Smirnova I. S., Bazhenov A. V., Shevchenko S. A., Kolyubakin A. I., Dilanian R. A. and Shekhtman V. Sh. (1996), "A cubic phase in ceramic Y-Ba-Cu-O doped with Nb," *Physica* C 267, 67.

23. Gloffke W., Skoczylas A., Ramanan A. and Wittingham M. S. (1991), "Effect of tungsten substitution in the $YBa_2Cu_3O_{7-x}$ system," *Mater. Lett.* 10, 477.

24. Hudakova N., Plechacek V., Dordor P., Flachbart K., Knizek K., Kovac J. and Reiffers M. (1995), "Influence of Pb concentration on microstructural and superconducting properties of BSCCO superconductors," *Supercond. Sci. Technol.* 8, 324.

25. Li Y. and Yang B. (1994), "Doping of the Bi-Sr-Ca-Cu-O system with V_B elements on $Bi_2Sr_2Ca_2Cu_3O_v$ phase formation," *J. Mater. Sci. Lett.* 13, 594.

26. Malo S., Pelloquin D., Maignan A., Michel C., Hervieu M. and Raveau B. (1996), "Superconducting oxycarbonates $Hg_{1-x}V_xSr_{4-v}Ba_vCu_2O_{7-\delta}CO_3$: three closely related structural families," *Physica* C 269, 1.

27. Salama K. and Lee D. F. (1994), "Progress in melt texturing of $YBa_2Cu_3O_x$ superconductor," *Supercond. Sci. Technol.* 7, 177.

28. Wong-Ng W. and Freiman S. W. (1994), "High-T_c superconducting Bi-Sr-Ca-Cu-O glass ceramics: a review," *Appl. Supercond.* 2, 163.

29. Alcock N. W. (1990), *Bonding and Structure: Structural Principles in Inorganic and Organic Chemistry,* Ellis Horwood Ed., New York.

30. Huvé M., Michel C., Maignan A., Hervieu M., Martin C. and Raveau B. (1993), "A 70 K superconductor: the oxycarbonate $Tl_{0.5}Pb_{0.5}Sr_4Cu_2(CO_3)O_7$," *Physica C* 205, 219.

CHAPTER 4

Impact Loading of Solid/Porous Media

4.1 NOTATION

A = Helmholtz free energy
A_f = area of flyer plate
C_P = heat capacity under constant pressure
C_V = heat capacity under constant volume
c_0 = sonic velocity at zero pressure
d = grain size
E = internal energy
E_F = energy component associated with fragmentation
E_k = kinetic energy per unit area of flyer plate
E_0 = initial internal energy
E_T = energy component associated with interface melting
E_{tot} = total energy
E_{vc} = void collapse energy
E_{vib} = vibration energy
f = parameter
G = strain energy release rate parameter
h = Planck's constant
h_e = thickness of explosive layer
h_f = thickness of flyer plate
J_c = critical current density
K = isothermal compressibility coefficient
K_{Ic} = critical stress intensity factor (Mode I)
k = Boltzmann's constant
L = melt fraction
m_e = mass of explosive

m_f = mass of flyer plate
m_p = mass of parent plate
m_{sc} = mass of superconducting material
P = shock-pressure
P_0 = ambient pressure
R = impact mass ratio
R_f = material fracture resistance
r = interatomic distance
r_c = radius of ceramic core
r_e = outer radius of explosive compaction setup
S = entropy
s = empirical constant
T = temperature
T_0 = ambient temperature
T_c = critical temperature
T_f = fusion temperature
t = time
t_c = cooling time
t_{dmin} = minimum pulse duration
t_s = solidification time
u = velocity of pulse
u_p = particle velocity
u_s = shock wave velocity
u_{sc} = shock wave velocity in superconducting material
V = volume
V_0 = theoretical volume of the solid
V_{00} = initial volume of powder
v_d = detonation velocity of explosive
v_p = initial impact velocity of flyer plate
Y = yield stress of solid material
Z = shock impedance
Z_f = impedance of flyer plate
Z_{sc} = impedance of superconducting material
Z' = impedance ratio (Z_f/Z_{sc})
α = initial setup angle
α_o = linear thermal expansion coefficient
β = collision angle
β_o = volumetric thermal expansion coefficient
γ = Grüneisen constant
$\Delta\alpha$ = increase of crack length
ΔH_f = fusion enthalpy
ΔM = magnetization difference
ΔT = transition temperature width

E = elastic modulus
ζ = explosive load ratio
θ = distention
θ_0 = initial distention
ν = frequency
ρ = density
ρ_0 = solid density
ρ_{00} = density of powder
ρ_e = bulk density of explosive
ρ_f = density of flyer plate
ρ_{sc} = mean density of superconducting material
$\varphi(\nu)$ = potential energy
χ' = real part of ac-susceptibility
χ'' = imaginary part of ac-susceptibility

4.2 GENERAL

In this chapter the dynamic powder consolidation is considered as for the fabrication of high-T_c superconducting components of various geometries, i.e. 3D-plates and axisymmetric billets, see References [1–3].

Dynamic processing is a very promising technique for compacting powdered materials in relative high density and in a near-net shape by means of chemical (explosion) or electromagnetic energy. High-strength structural components, such as hard ceramics (e.g., silicon carbide SiC) [4], special metals [5], amorphous materials [6], and also heat-sensitive materials, such as polymers [7] or polymer-based composites [8], which are difficult to be produced by conventional pressing and longtime sintering techniques, are very often manufactured by the rapid method of dynamic (explosive) compaction.

The short time of compaction and the confinement of high temperatures and pressures to prior particle boundary region make this method potentially attractive for advanced materials processing, such as for bulk oxide superconductors. Near-net-to-manufacture functional shapes, i.e., cylindrical billets or rods as wire precursors for current transport or composite conductors for high powder generators, can be fabricated by using the dynamic compaction technique.

The basic aspects of compaction technology, including the involved experimental techniques and the related process parameters, are also reported in product design and quality. Furthermore, the influence of the shock wave phenomena on the mechanical-structural integrity (soundness), microstructure, chemical transformations and the related superconducting properties of the compacted ceramics are presented in detail. Finally, a theoretical treatment based on the basic aspects of the explosive compaction thermohydrodynamics (Hugoniot curves, equations-of-state, etc.) is attended to formulate an analytical compaction model, consistent with experimental observations.

4.3 DYNAMIC CONSOLIDATION OF POWDERS

A variety of advanced materials are produced in powder form. Rapidly solidified metals, possessing a nanocrystalline, microdendritic or glassy structure, are synthesized and, similarly, many hard ceramic materials (diamond, BN, SiC, etc.) are produced in powder form. Recently, the new ceramic superconducting materials, with a critical temperature many degrees higher than that of the liquid nitrogen boiling point, are fabricated by various physico-chemical techniques, e.g., solid-state reaction, sol-gel, etc., in the form of powder. A method that can produce monoliths from powders, without changing their unique properties, would be very desirable.

Shock consolidation of powders is a one-stage densification/bonding process that can be useful for shaping the above-mentioned special and difficult-to-consolidate materials. Dynamic powder consolidation, which involves a very rapid and intense deposition of shock energy on powder particle surfaces, resulting in interparticle bonding, in the last 20 years, has been already applied in industry for the synthesis of diamond [9] and the wurtzite form of BN [10].

Shock wave powder consolidation is a technique that avoids prolonged heating and can be used for producing bulk solids in the form of disks, plates, cylinders, tubes and cones. The two basic mechanisms by which consolidation is achieved, as a shock wave passes through a porous medium, can be attributed to: (1) the plastic deformation and high-velocity impact of particles, resulting in filling the interstices, breaking down surface compound layers, heating and melting the particle near-surface material with subsequent welding; (2) the fracture of particles, resulting in filling the interstices, cleaning the surfaces, preferential heating of particle surfaces, leading, therefore, to partial melting and welding or solid-state diffusion bonding. The first mechanism seems to apply to the consolidation of metals, whereas the second one is more applicable to consolidating brittle solids, i.e., ceramics. Shock consolidation of engineering powders has been recently reviewed, see References [11,12].

A variety of processes can use the rapid energy deposition rates at the powder surface [12]:

- *Shock compaction:* Consolidation of the powder occurs because of the shock energy, preferentially being deposited on the particle surfaces, during the passage of shock waves.
- *Shock-enhanced sintering:* The shock-consolidated compact is statically compressed and normally sintered to produce the final product.

- *Shock conditioning:* The powder is shocked in any convenient geometry, remilled and conventionally sintered.
- *Shock-induced chemical synthesis:* The compound is formed from a powder mixture, during the passage of the shock wave, and at the same time consolidated.
- *Shock-induced phase transformations:* New structures with desirable properties can be formed under high pressure.
- *Chemically assisted shock consolidation:* It is a combination of shock-induced chemical synthesis and shock consolidation. Inert powders are mixed together with an exothermally reacting elemental mixture, whereas because of the passage of shock waves, a reaction between the elemental powders is induced, enhancing at the same time bonding between the initially inert and difficult-to-consolidate materials.

There are several means of depositing the energy required for dynamic consolidation on the powder-particle surfaces:

- Detonation of explosive is in direct contact with the powder or the powder container.
- Electromagnetic compaction is based on the dissipation of electromagnetic energy by an appropriate capacitor discharge in the compaction rig.
- Rapid deposition of energy occurs on the powder surface by high-amplitude pulsed laser (laser shock processing).
- Impact of projectiles against the powder or the powder container; velocities in the range of 200–3000 m/s are required to produce appropriate consolidation pressures. These projectiles can be accelerated electromagnetically, by compressed gases, by deflagration of gun powder or by detonation of explosives.

Of the different concepts introduced above, explosive compaction is perhaps the most advantageous technique for industrial net-shape manufacturing [13]. Different impact geometries can be fabricated by this technique. In the following sections, the main explosive compaction techniques are outlined.

High-T_c superconducting grains, produced by various physicochemical processes, contain a large number of superconducting phases, but, despite this, the current conducting ability of the bulk samples, compacted from these grains, is very low. The current carrying ability of the high-T_c superconducting materials mainly depends on their macro- and microstructure and, consequently, the current conduction of a superconducting material, with a given chemical composition, can be influenced by technological means. The bad current conduction of the high-T_c superconducting parts may be attributed to the contamination of the grain interfaces by fixed or adherent gases or, more usually, by normal (insulating) phases, produced during processing, resulting, therefore, in the interruption of the continuous current conduction. A suggested solution to these problems is the creation of new clean grain boundaries and new large contact surfaces by breaking the original grains and bringing these new surfaces into very close contact with each other.

Explosive powder compaction makes possible the realization of it, because shock waves, originated from explosive detonation and propagated through the porous media, can create shock pressures, ranging from about one to a few hundreds GPa and local temperature jumps of some thousands degrees in the time frame of 1–10 μs; they result in fracturing the original grains and they also help in sintering. Note that more dense material structures are obtained with high shock pressures than with conventional static compaction processes. At microstructural level, the compacted solid contains a variety of primarily line defects (dislocations) that would provide flux pinning centers in Type II superconductors [14].

Several attempts have been made to consolidate cuprate superconducting powders and specially the Y-Ba-Cu-O compound, applying various shock wave techniques. Several defects at macro- and microscale were observed: large scale longitudinal and transverse cracking [15]; incompleteness and disorder in crystallinity, generated by the high shock pressures [16]; catastrophic structural changes deteriorating the residual superconductivity, such as precipitate-like defect clusters, associated with the shock-induced large strain fields, and the transformation of the orthorhombic (superconducting) phase to tetragonal (normal) phase due to the oxygen loss that resulted from the elevated temperature in the high-pressure state [17].

Explosive compaction is followed by conventional plastic deformation processing, e.g., rolling, extrusion, wire-drawing, etc., to fabricate metal-sheathed superconducting composites of various geometries, strips, rods, wires, etc. Several superconducting powders are used: K-doped ceramics of the Y-Ba-Cu-O system and Pb-doped Bi-Sr-Ca-Cu-O compounds. Applications of these high-T_c superconducting composite materials can be found in the electrical and electronic industry, e.g., in the construction of levitating bearings for measuring equipment and of devices of high numbers of revolution; one part of these bearings, the stator or the rotor, is made of permanent magnet and the other of high-T_c superconducting material. Note that the levitating effect is based on the Meissner effect.

4.3.1 Explosive Compaction

RECTANGULAR 3D CONFIGURATION

This experimental setup is used for shaping rectangular shaped components and is a combination between explosive welding and compaction, see Figure 4.1. The fabrication of rectangular components, such as copper compacts [18] and

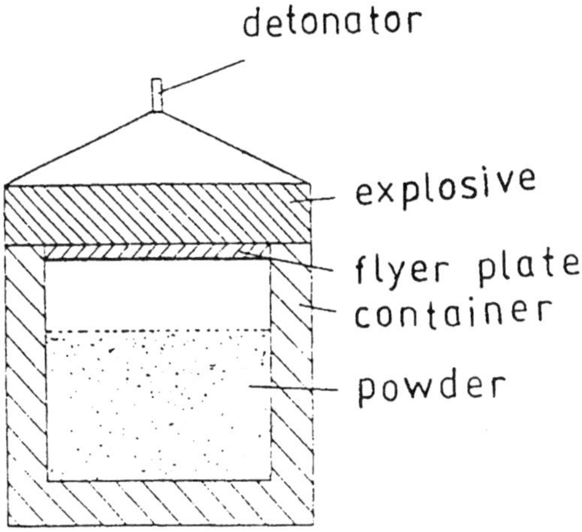

FIGURE 4.1. Experimental setup of the rectangular cladding/compaction system.

silver-sheathed high-T_c superconductors [2,3] was performed by using such a 3D configuration setup. The explosive welding technique was first used to fabricate cladded parts from dissimilar metals, such as titanium/steel, copper/aluminium, etc., after the Second World War [19].

The powder is inserted and compressed in the grooves of the lower metal plate (parent plate), whereas the explosive layer is placed on the upper plate (flyer plate), which is separated from the parent plate by a gap, known as stand-off distance. When the explosive is charged, the flyer plate is accelerated against the parent plate, generating very high impact pressures, of the order of 10 GPa, leading to powder compaction and to welding between the two metallic plates. To avoid the defects induced by the shock wave instabilities, a buffer material is placed at the edges of the plate and at points of superposition of stress waves. High-velocity impact leads, very often, to metal jetting and the transition from the laminar to turbulent flow, resulting in the formation of von Kármán "vortices" between the two metal interfaces. Note that wavy interfaces are a very common characteristic of the explosively cladded components.

CYLINDRICAL SINGLE-TUBE CONFIGURATION

This is the most common configuration used for shock compaction [20]. The main parts of this setup are shown in Figure 4.2(a). A similar compaction rig was used for the consolidation of YBCO superconducting powders [21]. The shock pressure on the powder can vary according to the variation of the amount and the type of the explosive used. By increasing the detonation velocity, from 3500 m/s (the lowest range of ammonium nitrate and fuel oil [ANFO] explosives) to 7000 m/s (for plastic explosives, like PETN), the peak shock pressure

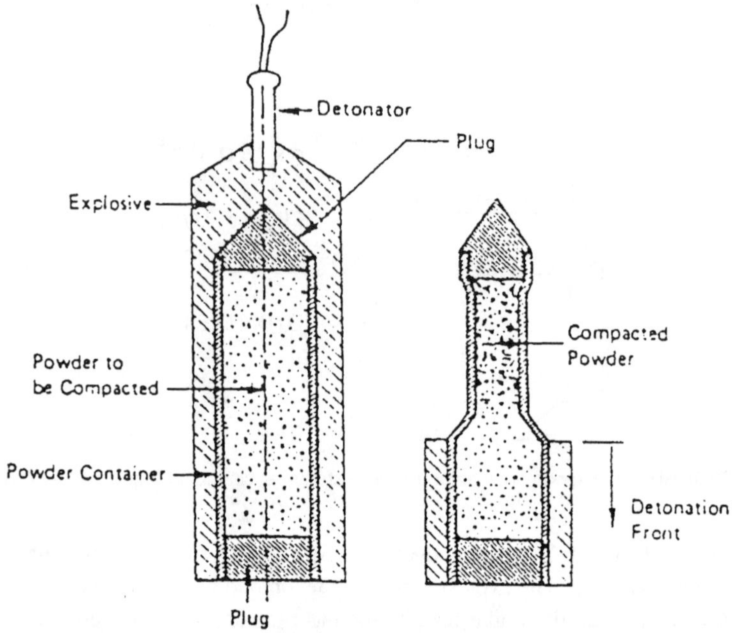

FIGURE 4.2. Configuration of a cylindrical single-tube compaction.

is four times higher, because the pressure varies with the square of the detonation velocity. The pressure pulse converges toward the central axis of the cylinder and, if excessive, a hole is generated along the cylinder axis, known as Mach stem. The Mach stem can be eliminated by putting a solid metal rod (mandrel) along the axis or by adjusting the shock and detonation conditions. This modified setup has been used for superconducting powder compaction [22].

The single-tube configuration is generally used for compacting mainly soft and ductile powders, with hardness lower than 500 kg/mm^2. The superconducting ceramics are not hard enough, like other engineering ceramics, such as carbides, nitrides, alumina, etc., satisfying, therefore, this condition. Finally, to avoid end defects due to the shock wave instability, a buffer material, usually cheap MgO powder, is placed near the top and bottom ends of the cylindrical rig.

CYLINDRICAL DOUBLE-TUBE CONFIGURATION

This technique has been developed for the shock synthesis of diamond and afterward is used for powder consolidation purposes [13]. The powder is contained in the internal tube, whereas the external tube is surrounded by the explosive charge, which is detonated at one end; the external tube acts as the flyer tube, impacting the internal tube. This technique generates pressures in the powder that can be several times higher than the ones generated by the single-tube technique. The main advantage of this technique is that it allows for the use of low detonation-velocity explosives for consolidating hard powders, therefore, avoiding or minimizing cracking of the compacts. Significant improvement in compact quality have been obtained in Ni-base superalloys, Ti alloys and Al-Li alloys. The basic experimental setup is shown in Figure 4.3, which is similar to the corresponding single-tube setup.

HIGH-TEMPERATURE SETUP

When excessive cracking is present after shock consolidation at ambient temperature, sound components may be often obtained by preheating the powders. The high temperature can induce additional ductility to the powders and

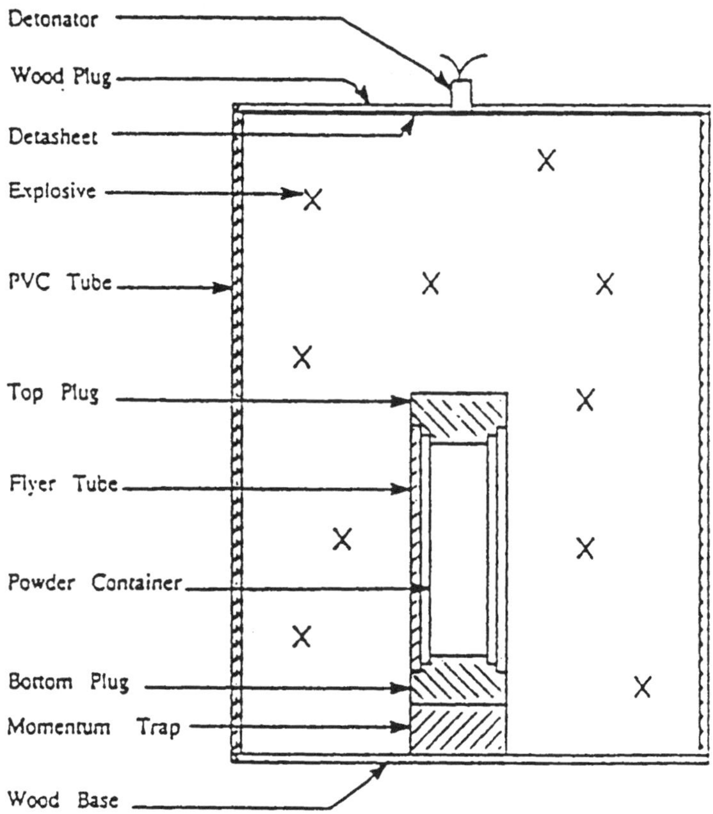

FIGURE 4.3. A schematic diagram of the experimental setup for the cylindrical axisymmetric double-tube compaction.

reduce, therefore, their strength and hardness. Another advantage of high temperatures is that the shock energy required to melt the powder surfaces is decreased. Systematic research work on high-temperature powder compaction can be found in References [23,24].

4.3.2 Electromagnetic Compaction

A schematic diagram of the electromagnetic compaction setup, shown in Figure 4.4, has been used for the consolidation of YBCO powders, see Reference [25]. The powder to be consolidated is placed between a thin-walled metal tube (Ag 999). Very often, a steel mandrel is placed in the middle of

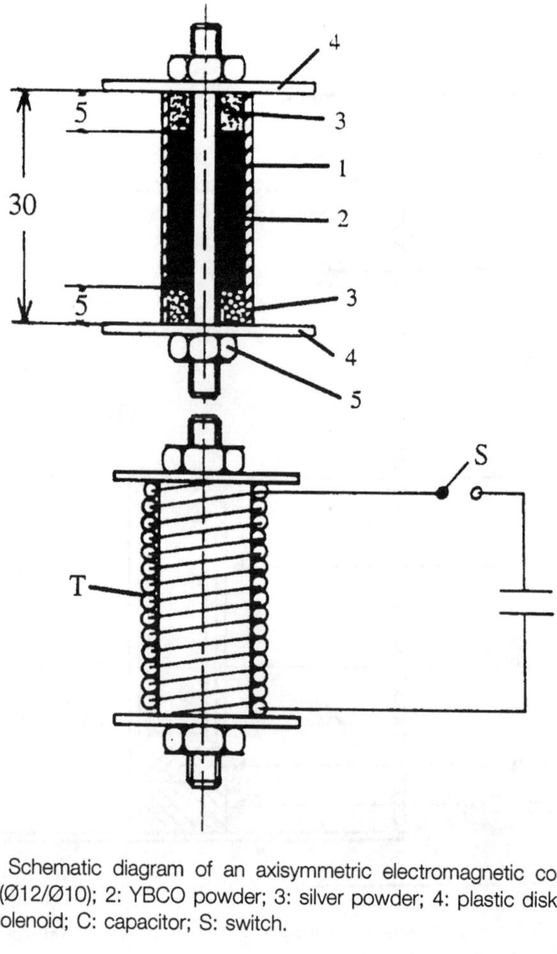

FIGURE 4.4. Schematic diagram of an axisymmetric electromagnetic compaction rig. 1: silver tube (Ø12/Ø10); 2: YBCO powder; 3: silver powder; 4: plastic disk; 5: steel bolt (AISI M5); T: solenoid; C: capacitor; S: switch.

the container, to avoid central cracking and hole formation. Silver powder (Ag 999) was used at the ends of the tube for encapsulating the ceramic powder. The whole container is inserted into an electromagnetic cylindrical coil, consisting of a certain number of windings made of insulated copper wire. The strong transient magnetic field, developed in the coil (T), due to the discharge of the capacitor bank (C) with a stored electrical energy through switching-off (S), induces eddy currents in the highly conductive silver tube, which, subsequently, lead to the development of another magnetic field. The interaction of these two magnetic fields results in the collapse of the silver tube and to the subsequent consolidation of the powder, i.e., the tube acts as a tubular punch. Extensive work on the electromagnetic compaction process of metallic materials can be found in Reference [26].

4.4 PROPAGATION OF SHOCK WAVES: THEORETICAL MODELING

4.4.1 Basic Notions

HUGONIOT CURVES

Shock waves are characterized by a steep front and require a state of uniaxial strain, which allows for the development of very high hydrostatic stress component. The calculation of the shock wave parameters is based on the Rankine-Hugoniot conservation equations. The theoretical treatment, attributed to Rankine-Hugoniot [27], is called the *hydrodynamic approach.*

When a material is under dynamic (or shock) conditions the following assumptions are valid according to the above-mentioned hydrodynamic treatment:

- A shock is a discontinuous surface and has no apparent thickness.
- The shear modulus of the material is assumed to be zero, so that it responds to the wave as a fluid; hence, the theory is restricted to higher pressures.
- Body forces, such as gravitational, and heat conduction at the shock front are negligible.
- There is no elastoplastic behavior.
- Material does not undergo phase transformations.

The basic mathematical expression for the establishment of a shock wave is that the velocity of the pulse, u increases with increasing pressure, P that means: $(\partial P/\partial u) > 0$ and $(\partial^2 P/\partial u^2) > 0$, as $P\uparrow$, $u\uparrow$. On the basis of these assumptions, the travel of the shock wave into a solid can be simulated to the adiabatic gas compression through a moving piston, see Figure 4.5. At the beginning, i.e., $t = 0$, the piston does not move and the gas is under pressure, P_0 and initial density, ρ_0, see Figure 4.5(a). After a certain time interval, the piston moves, pressing the gas with a speed up to u_p, see Figure 4.5(b). The disturbance of the gas density, i.e., shock wave, caused by the piston motion, travels by a speed u_s ahead of the piston, see Figure 4.5(b) and (c). The compressed region has a pressure, P and a density, ρ. The basic equations that are extracted from the conservation principles applied on both sides of the shock wave interface are the following:

(1) Conservation of mass

$$\rho_0 u_s = \rho(u_s - u_p) \tag{4.1}$$

(2) Conservation of momentum

$$P - P_0 = \rho_0 u_s u_p = Z u_p \tag{4.2}$$

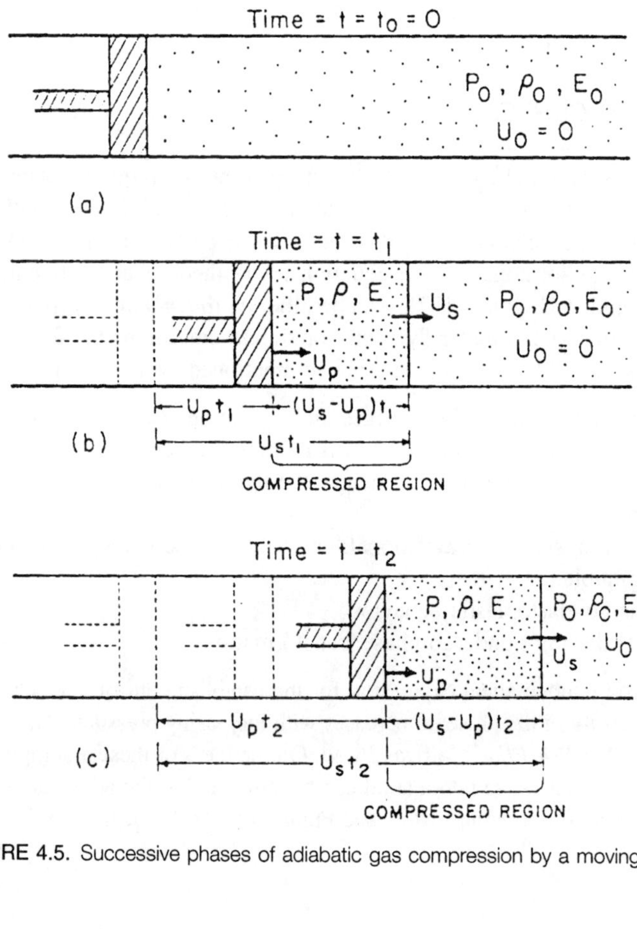

FIGURE 4.5. Successive phases of adiabatic gas compression by a moving piston.

(3) Conservation of energy

$$E - E_0 = \frac{1}{2}(P + P_0)(V_0 - V) \tag{4.3}$$

An additional equation is needed to determine the relation between the shock wave velocity, u_s and the particle velocity, u_p. This equation, known as *equation-of-state (EOS)*, is experimentally defined and it has usually a polynomial form. The simplest expression is linear, as:

$$u_s = c_0 + su_p \tag{4.4}$$

where c_0 is the sonic velocity in the material at zero pressure and s an empirical constant.

It may be noted that if the material has a porosity or undergoes a phase transformation, the linear equation of state is no longer valid and has to be modified. Equation (4.3), derived from the conservation of energy, establishes a relation between P and ρ, immediately behind the shock front. This pressure-density relationship is usually known as the *Rankine-Hugoniot* equation or simply the "Hugoniot" and, in graphic form, it is shown in Figure 4.6 as a

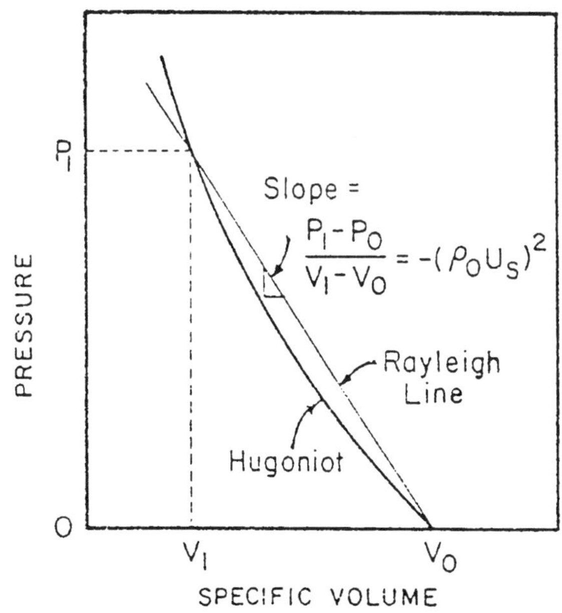

FIGURE 4.6. Characteristic P-V Hugoniot curve showing the Rayleigh line.

P-V curve, where *P* is the shock pressure and *V* the volume. Hence, a Hugoniot curve may be defined as the locus of all shocked states in a material and, essentially, it describes the material properties. The straight line, joining the points (P_0, V_0) and (*P*, *V*) in the curve, is known as the *Rayleigh line*. Note that when the pressure increases, from its initial value P_0, it does not necessarily follow the *P-V* path, but it changes rather discontinuously from its initial value to the final one.

SHOCK WAVE PROFILE

The ideal shock wave profile consists of a sharp front, which is called *shock front,* and a part of *pressure release,* see 2 and 4 of Figure 4.7, respectively. The part of the shock wave, preceded of the whole stress pulse, is an elastic wave characterized by an intensity equal to the elastic limit of the material, see 1 of Figure 4.7. The pressure of the elastic wave is called *Hugoniot Elastic Limit (HEL).* Note that for extremely high shock pressures the shock wave overcomes the elastic precursor.

The shock wave, developed by the high-speed impact of a projectile impinging onto a target, has a trapezoidal form, see Figure 4.7(a). The length of the shock front, see 3 of Figure 4.7(a), determining the pulse duration, depends on the time required for the wave to travel through the matter. When the shock waves are formed, by the detonation of an explosive substance in contact to the material or, by pulsed laser radiation, the form of the pulse is triangular as shown in Figure 4.7(b). Region 4 of Figure 4.7 is a pressure relaxation period, often called *rarefaction.* A real shock wave profile (*P-t*), determined by laser interferometry, is shown in Figure 4.8.

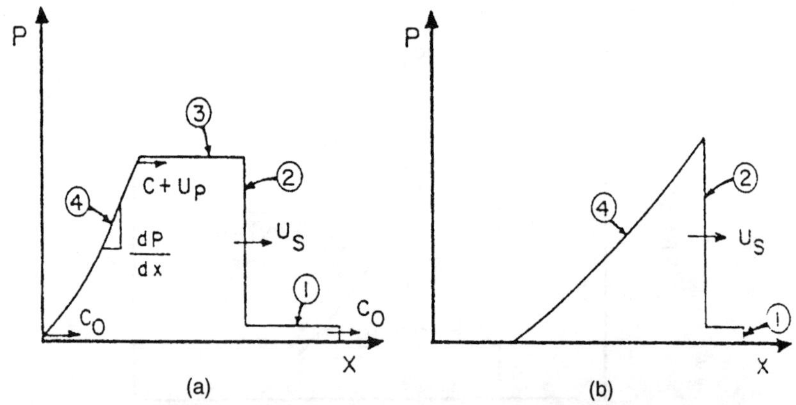

FIGURE 4.7. Idealized shock wave profiles formed because of (a) impact of a projectile (trapezoidal pulse) and (b) pulsed laser radiation (triangular pulse). 1: elastic wave; 2: shock front; 3: shock length; 4: rarefaction.

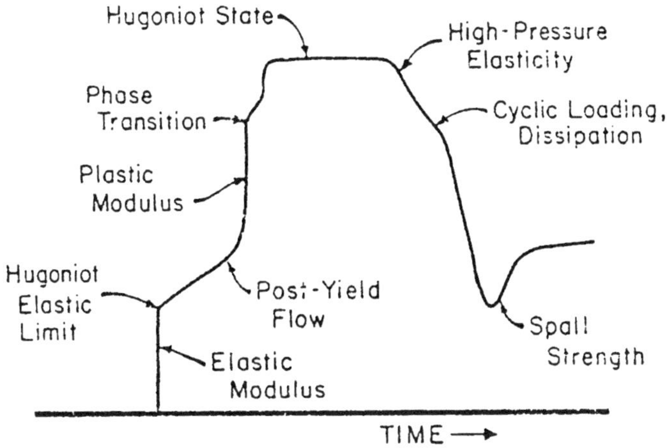

FIGURE 4.8. Real shock wave profile formed because of impact.

SHOCK-WAVE REFLECTION AND TRANSMISSION

When a shock wave propagates through a medium A enters a medium B, changes in pressure, wave velocity and density do occur. It is convenient to define the *shock impedance, Z,* as the product of the initial density, ρ_0 and the shock wave velocity, u_s. This can be approximated as the product of initial density and sonic velocity at zero pressure, c_0. The impedance shows the highest values for materials with high densities and high sonic velocities. In high-impedance materials, i.e., solids or porous, the best way to treat the transfer of the way from medium A to medium B is by means of the impedance-matching technique. Continuity at the A/B boundary indicates that the particle velocity and pressure will be the same for both materials, see Reference [12].

Transmission of Shock Wave from Material A to Material B when $Z_A < Z_B$

Figure 4.9(a) shows the pressure-particle velocity for materials A and B, which are encountered as semi-infinite media. The slope of the dashed line represents the impedance of the material A. At the interface, the pressure P_1 will change, so that equilibrium can be reached, as obtained by the impedance-matching method shown in Figure 4.9(a). The AR curve corresponds to the reflected pulse, which intersects the B-curve at P_2, which is the pressure developed in medium B. The sequence of the pressure profiles is shown in Figure 4.9(b). As the shock front reaches the A/B interface, the pressure rises

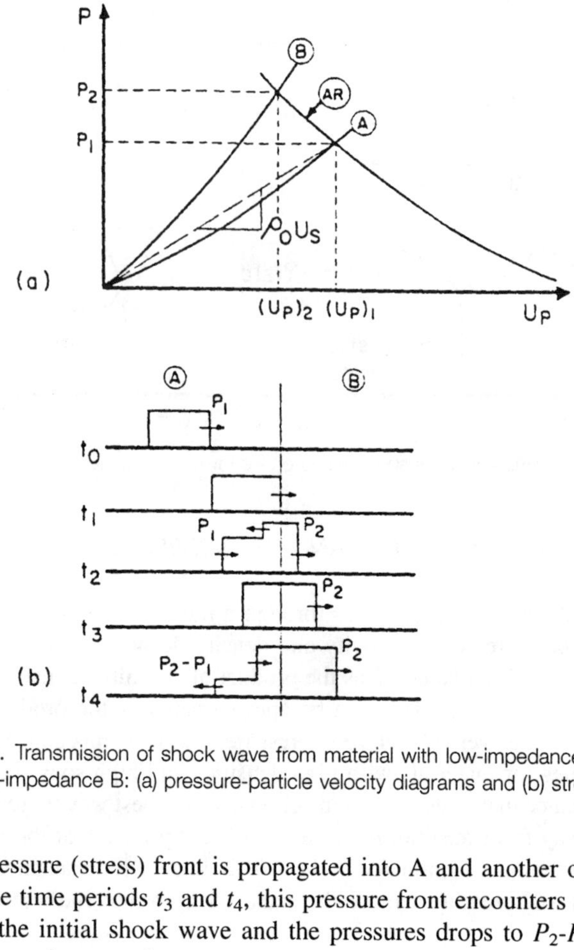

FIGURE 4.9. Transmission of shock wave from material with low-impedance A to material with high-impedance B: (a) pressure-particle velocity diagrams and (b) stress profiles.

to P_2. A pressure (stress) front is propagated into A and another one into B. Between the time periods t_3 and t_4, this pressure front encounters the release portion of the initial shock wave and the pressures drops to P_2-P_1; this reduced pressure front continues to propagate at the left, inside material A. The particle velocity at the high-pressure region is denoted as $(u_p)_2$ and, thus, the continuity is maintained.

Transmission of Shock Wave from Material A *to Material* B *when* $Z_A > Z_B$

The impedance-matching technique is illustrated in Figure 4.10(a). Now, the pressure P_2 is lower than the P_1. This pressure will produce a release pulse through the material A, which travels freely until it encounters the primary pulse; subsequently, a tensile stress wave is developed and propagated through the material B, see Figure 4.10(b). If the amplitude of this tensile pulse is suf-

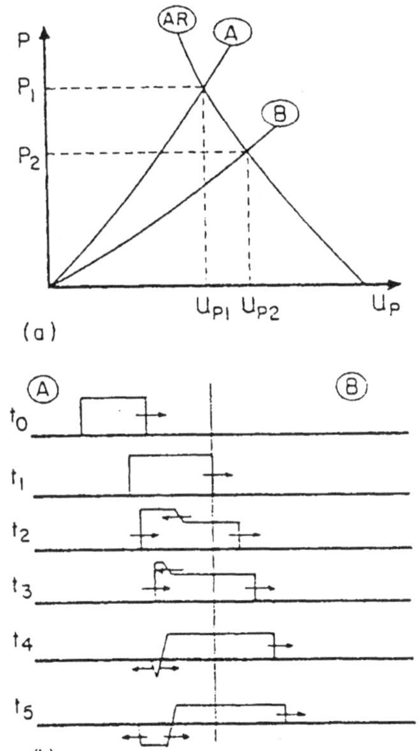

FIGURE 4.10. Transmission of shock wave from material with high-impedance A to material with low-impedance B: (a) pressure-particle velocity diagrams and (b) stress profiles.

ficient enough, exceeding the tensile strength of the material, a spall may be formed.

4.4.2 Equations-of-State (EOS)

FUNDAMENTAL APPROACH

The theoretical calculation of the shock parameters of materials, for the determination of the equations-of-state (EOS), requires the basic knowledge of quantum and statistical mechanics [12]. The compression of atoms in a crystal leads to interpenetration and overlapping of the electronic shells, resulting in the creation of very high short-range forces. The potential energy-interatomic distance curve describes the energy and force between two atoms as a function of their separation, see Figure 4.11(a). For ionic materials, the attractive forces vary with $1/r^2$ (Coulomb forces), whereas the repulsive forces vary with $1/r^4$, see Figure 4.11(b). The minimum energy point defines the minimum distance between the atoms, being in equilibrium condition. When atoms are in a very close distance, the electronic atoms are in a strong interaction, leading to very high repulsive forces. By increasing the interatomic distance, the attractive force increases. The inflection point in the energy curve determines the maximum attractive force point and, therefore, from the force curve the pressure-distance curve, i.e., the compressibility curve, can be derived, see Figure 4.11(c). The shape of the compressibility curve indicates that the slope increases with increasing pressure. From this "cold compression" curve the shock-Hugoniot curve can be calculated, as in the next stage.

Bridgman established compressibility curves for static high pressures for a great variety of materials [28]. However, these curves are not available for

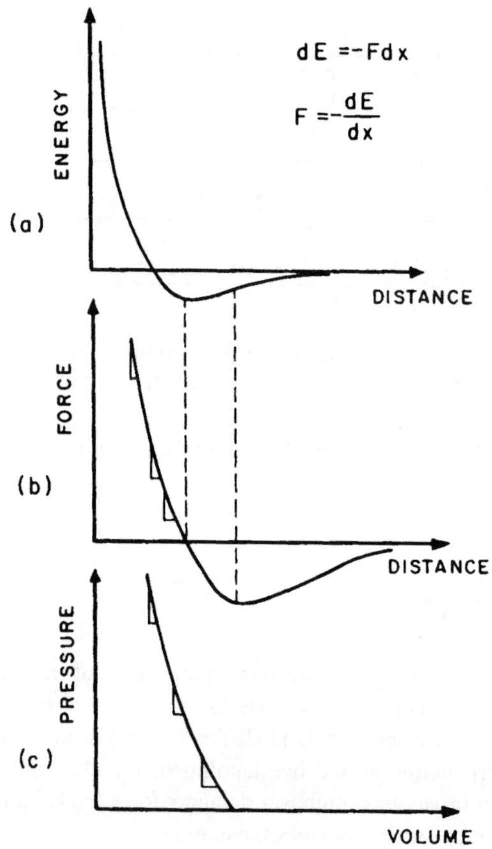

FIGURE 4.11. Distribution of (a) energy, (b) force and (c) pressure as a function of interatomic distance.

very high pressures, i.e., greater than 10 GPa. To obtain the Hugoniot curves in this range, it was not possible to achieve these very high pressures statically, but only by dynamic loading techniques. For this reason the Mie-Grüneisen EOS may be applicable for the determination of the shock parameters; the Grüneisen constant, γ, originates from statistical mechanics.

According to statistical mechanics, atoms are considered as quantized harmonic oscillators, their energy of the *n*th level, being equal to $nh\nu$, where h is the Planck's constant and ν is the frequency of the vibration (the ground state $h\nu/2$ is not included). There are three discrete frequencies of vibration per atom. For a system containing N atoms, there are $3N$ different frequencies (ν_i, $i = 1,2,...,3N$), whereas the mean total vibration energy of $3N$ different oscillators may be expressed by the following formula:

$$E_{vib} = \sum\left[\frac{h\nu_i}{2} + \frac{h\nu_i}{(e^{h\nu_i/kT} - 1)}\right] \tag{4.6}$$

The total energy, E_{tot} is a sum of the potential energy, $\varphi(\nu)$ and the total vibration energy, E_{vib}, i.e.,

$$E_{tot} = \varphi(\nu) + \sum\left[\frac{h\nu_i}{2} + \frac{h\nu_i}{(e^{h\nu_i/kT} - 1)}\right] \tag{4.7}$$

The *Helmholtz free energy, A,* derived from Equation (4.7) is

$$A = \varphi(\nu) + \sum\left(\frac{h\nu_i}{2}\right) + kT\sum[\ln(1 - e^{-h\nu_i/kT})] \tag{4.8}$$

The pressure is calculated as a partial derivative of Helmholtz free energy, with respect to volume at constant temperature

$$P = -\left(\frac{\partial A}{\partial V}\right)_T = -\frac{d\varphi}{dV} + \frac{1}{V}\sum\gamma_i\left[\frac{h\nu_i}{2} + \frac{h\nu_i}{(e^{h\nu_i/kT} - 1)}\right] \tag{4.9}$$

The parameter γ_i is defined as

$$\gamma_i = -\frac{V}{\nu_i}\left(\frac{\partial \nu_i}{\partial V}\right)_T = -\left(\frac{\partial \ln \nu_i}{\partial \ln V}\right)_T \tag{4.10}$$

Assuming that all the oscillators possess the same Grüneisen constant, i.e.,

$$\gamma = -\left(\frac{\partial \ln \nu}{\partial \ln V}\right)_T \tag{4.11}$$

If $\gamma_i = \gamma$ = const, combining with Equation (4.9), it yields

$$P = -\frac{d\varphi}{dV} + \frac{\gamma}{V} E_{vib} \tag{4.12}$$

Applying this equation at the absolute zero temperature, 0 K,

$$P_{0K} = -\frac{d\varphi}{dV} + \frac{\gamma}{V} E_{vib,0K} \tag{4.13}$$

and by subtracting Equation (4.13) from Equation (4.12)

$$P - P_{0K} = \frac{\gamma}{V}(E - E_{0K}) \tag{4.14}$$

The latter Equation (4.14) represents the Mie-Grüneisen EOS, which relates a (P,V,E) state with the pressure and internal energy at 0 K, and it can also be referred to the Hugoniot plot (P_H, V_H) as

$$P - P_H = \frac{\gamma}{V}(E - E_H) \tag{4.15}$$

From Equation (4.15) the Grüneisen parameter can be expressed as

$$\gamma = V\left(\frac{\partial P}{\partial E}\right)_V \tag{4.16}$$

From the thermodynamic relations

$$\gamma = V\left(\frac{\partial P}{\partial E}\right)_V = V\left(\frac{\partial P}{\partial T}\right)_V \left(\frac{\partial T}{\partial E}\right)_V = \frac{V}{C_V}\left(\frac{\partial P}{\partial T}\right)_V$$

or, finally

$$\gamma = -\frac{V}{C_V}\left(\frac{\partial P}{\partial V}\right)_T \left(\frac{\partial V}{\partial T}\right)_P \tag{4.17}$$

where $(1/V)(\partial V/\partial T)_P = \beta_o$ is the volumetric thermal expansion coefficient and $-(1/V)(\partial V/\partial P)_T = K$ the isothermal compressibility coefficient. Taking also into account that $\beta_o = 3\alpha_o$ (α_o is the linear thermal expansion coefficient), the following expression may be obtained

$$\frac{\gamma}{V} = \frac{3\alpha_o}{K \cdot C_V} = const \tag{4.18}$$

Note that, for practical applications the Grüneisen parameter can be approximated as, see also Equation (4.4)

$$\gamma \approx 2s - 1 \tag{4.19}$$

Equation (4.19) is very useful for extracting the Grüneisen parameter for unknown materials. Note, however, that for the majority of metallic materials, the γ parameter lies between 1 and 3.

EOS FOR SOLID AND POROUS MATERIALS

On the basis of the fundamental Grüneisen EOS, given by Equation (4.15), the principal equations-of-state for solid and for porous materials can be deduced. The three basic equations, i.e., the Rankine-Hugoniot relationships for the conservation of mass, momentum and energy, see Equations (4.1)–(4.3), are applied for the solid and porous materials. Note that the subscript H is used to describe the corresponding solid parameters as

(1) For the solid

$$\rho_0 u_{sH} = \rho(u_{sH} - u_{pH}) \tag{4.20}$$

$$P_H = \rho_0 u_{sH} u_{pH} \tag{4.21}$$

$$E_H = \frac{1}{2} P_H (V_0 - V) \tag{4.22}$$

$$u_{sH} = c_0 + s u_{pH} \tag{4.23}$$

Equation (4.21) can be expressed as

$$P_H = c_0^2 \frac{V_0 - V}{[V_0 - s(V_0 - V)]^2} \tag{4.24}$$

(2) For the porous material

$$\rho_{00} u_s = \rho(u_s - u_p) \tag{4.25}$$

$$P = \rho_{00} u_s u_p \tag{4.26}$$

$$E = \frac{1}{2}P(V_{00} - V) \tag{4.27}$$

By substituting Equations (4.22), (4.24) and (4.25) into Equation (4.15), the following EOS for the powder can be derived:

$$P = \frac{c_0^2(V_0 - V)[2V - \gamma(V_0 - V)]}{[2V - \gamma(V_{00} - V)][V_0 - s(V_0 - V)]^2} \tag{4.28}$$

For an equation, expressing the particle velocity as a function of pressure, the following relationship may be obtained from Equations (4.25) and (4.26):

$$u_p = \sqrt{P(V_{00} - V)} \tag{4.29}$$

Combining Equations (4.24) and (4.28), the ratio P/P_H may be obtained as

$$\frac{P}{P_H} = \frac{2V - \gamma(V_0 - V)}{2V - \gamma(V_{00} - V)} > 1 \tag{4.30}$$

The above relationship, Equation (4.30), indicates that the Hugoniot curves for porous materials are lying in the right side of the corresponding *solid Hugoniots*. Increasing the porosity ($V_{00}\uparrow$), see Figure 4.12, the *porous Hugoniots* may be obtained (right part of the curve). The energy dissipated by the shock wave in the powder is a very important parameter. From the definition of en-

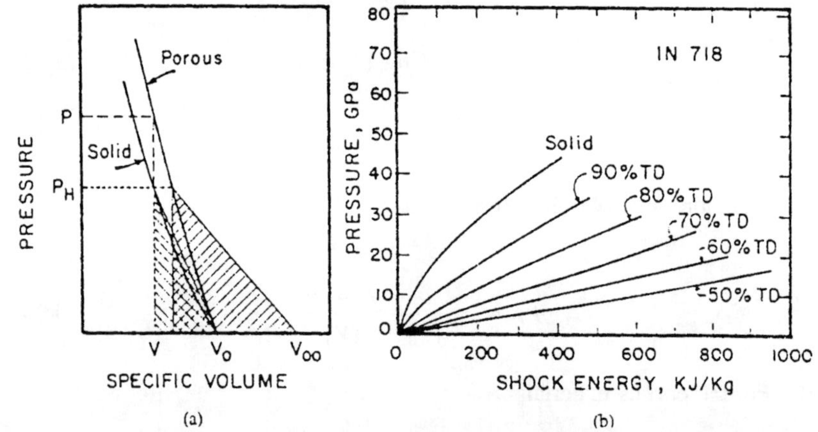

FIGURE 4.12. (a) Schematic P-V Hugoniot curves for solid and porous material. (b) Calculated P-E curve for solid and powders with various densities for INCONEL 718 (from Reference [12]).

ergy, see also Rankine-Hugoniot Equations (4.22) and (4.27), the energy at a certain pressure level P_H, for both solid and porous material, is equal to the shaded areas shown in the curve of Figure 4.12(a). The energy dissipated in the powder material, corresponding to the same pressure level, is higher than that dissipated into the solid material. Increasing the initial porosity the energy increases, see also Figure 4.12(b).

EXPERIMENTAL DETERMINATION OF EOS

The experimental determination of the EOS of an unknown material is based on the so-called *impedance-matching* technique. This technique is used when the EOS impacting material (projectile) are known, whereas the target's EOS have to be defined, see Figure 4.13(a) and Reference [12]. The measurements of the shock and the impact velocity can be obtained by measuring the surface velocity by laser interferometry. Figure 4.13(a) shows a gas-gun experiment; three pins in the barrel establish the velocity of the projectile. Figure 4.13(b) shows the Hugoniot for the projectile, with the known EOS, corresponding to the impact velocity, v_1. By measuring the shock wave velocity, u_s, and initial density of the unknown material, the Rayleigh line with a slope $\rho_0 u_s$ can be defined; this line intersects the known EOS at point 1. By repeating the procedure for different impact velocities, points 2 and 3 of the shock Hugoniot curve can be also determined.

4.4.3 Thermodynamic Treatment

The differential change of the internal energy can be approximated in terms of the thermodynamic parameters, entropy, S and volume, V, as

$$dE = T \cdot dS - P \cdot dV \tag{4.31}$$

Expressing the entropy, S as a function of its variables, T and V, i.e., $S = S(T,V)$, and differentiating Equation (4.31), it can be obtained

$$dS = \left(\frac{\partial S}{\partial T}\right)_V dT + \left(\frac{\partial S}{\partial V}\right)_T dV$$

or, finally

$$T \cdot dS = T\left(\frac{\partial S}{\partial T}\right)_V dT + T\left(\frac{\partial S}{\partial V}\right)_T dV \tag{4.32}$$

The heat capacity under constant volume, C_V is defined as

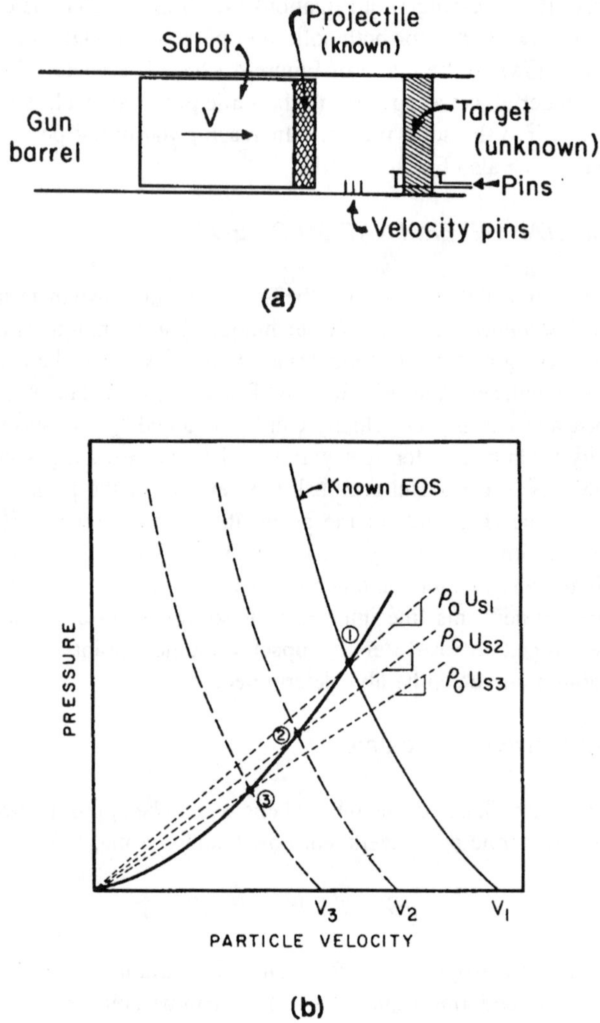

FIGURE 4.13. (a) Schematic illustration of a gas-gun barrel experimental setup. (b) Impedance-matching technique for the calculation of the shock-Hugoniot curve (from Reference [12]).

$$C_V = \left(\frac{\partial E}{\partial T}\right)_V = \left(\frac{\partial E}{\partial S}\right)_V \left(\frac{\partial S}{\partial T}\right)_V = T\left(\frac{\partial S}{\partial T}\right)_V \tag{4.33}$$

According to the chemical thermodynamics [29], related to the Maxwell equations, identities between the second order mixed partial derivatives of the thermodynamic potentials are apparent and, therefore, applying the Maxwell equations to the Helmholtz free energy, the following expression is obtained:

$$\left(\frac{\partial^2 A}{\partial V \cdot \partial T}\right)_V = \left(\frac{\partial^2 A}{\partial T \cdot \partial V}\right)_T$$

or

$$\left[\frac{\partial}{\partial T}\left(\frac{\partial A}{\partial V}\right)\right]_V = \left[\frac{\partial}{\partial V}\left(\frac{\partial A}{\partial T}\right)\right]_T$$

or, finally

$$\left(\frac{\partial P}{\partial T}\right)_V = \left(\frac{\partial S}{\partial V}\right)_T \tag{4.34}$$

Combining Equations (4.31)–(4.34), the following expression is derived:

$$dE = C_V dT + T\left(\frac{\partial P}{\partial T}\right)_V dV - P \cdot dV \tag{4.35}$$

Following Equation (4.16), in conjunction with the Grüneisen relationship, it yields to

$$\left(\frac{\partial P}{\partial T}\right)_V = \left(\frac{\partial P}{\partial E}\right)_V \left(\frac{\partial E}{\partial T}\right)_V = \gamma \frac{C_V}{V} \tag{4.36}$$

Substituting from Equation (4.36) into Equation (4.35), the following expression may be derived

$$dE = C_V dT + \left(T\frac{\gamma}{V} C_V - P\right) dV \tag{4.37}$$

Differentiating the basic Rankine-Hugoniot relationship, i.e., Equation (4.27), and substituting the derivative $(dE/dV)_H$ from the Equation (4.37), it yields to

$$C_V\left(\frac{dT}{dV}\right)_H + T\frac{\gamma}{V} C_V = \frac{1}{2}\left(\frac{dP}{dV}\right)_H (V_{00} - V) + P \tag{4.38}$$

Evaluating the derivative $(dP/dV)_H$ from the Hugoniot plot and introducing it into the differential Equation (4.38), the following expression is derived for temperature in terms of P, V and V_{00}:

$$T = T_0 e^{\frac{\gamma_0}{V_0}(V_{00}-V)} = (V_{00} - V)\frac{P}{2C_V} + \frac{1}{2C_V} e^{-\frac{\gamma_0 V}{V_0}} \int P e^{\frac{\gamma_0 V}{V_0}} \left[2 - \frac{\gamma_0 (V_{00} - V)}{V_0} \right] dV \tag{4.39}$$

Subsequently, the temperature, induced by shock loading into the material, can be calculated from Equation (4.39), see also Reference [12].

4.4.4 Shock-Induced Energy

In Figure 4.14 the basic features of physical and chemical energetic processes induced by dynamic powder consolidation are presented, see also Reference [12]. The thermodynamic formulation of the compaction models requires the determination of the total shock energy distribution to these

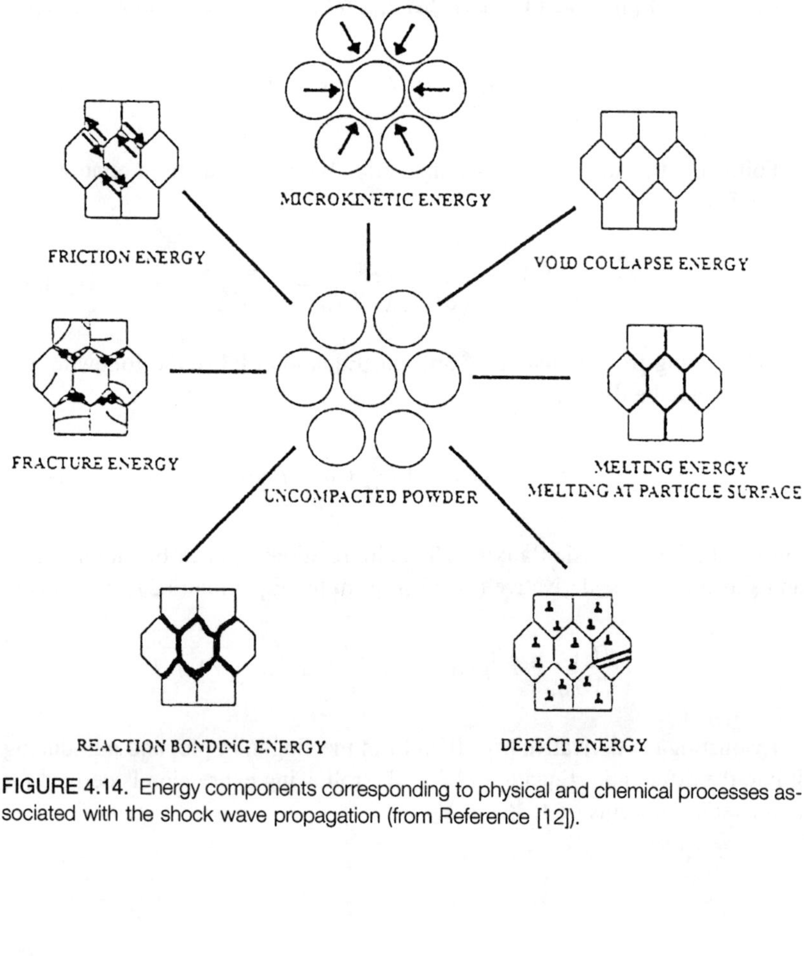

FIGURE 4.14. Energy components corresponding to physical and chemical processes associated with the shock wave propagation (from Reference [12]).

processes. The energy components, corresponding to the physical and chemical processes, may be classified as the following, see Figure 4.14:

- *Microkinetic energy:* It deals with the kinetic energy of the particles due to the shock wave passage through the powder.
- *Void collapse energy:* This energy component is related to the plastic or viscoplastic deformation of the powder particles, leading to the effective closing of the interparticle voids.
- *Melting energy at particle surface:* This energy component corresponds to the heat dissipated because of the shock wave propagation at the surface region of the powder particles; melting of a thin particle surface layer very often results from shock wave heating.
- *Defect energy:* It corresponds to the energy of formation of various microstructural defects, such as dislocations, stacking fault, twins, etc.
- *Chemical reaction energy:* It deals with the formation of various phases and compounds, induced by shock pressure and temperature; these phases are usually located at the near particle interface region.
- *Fragmentation or fracture energy:* This energy component is required for the consolidation of brittle powders, such as ceramics. Shock loading leads to intense fracturing of the initial grains and stacking, leading to void decrease and to the consolidation of the porous material.
- *Friction energy:* The heat released is due to the friction developed at the particle-particle contacts.

4.4.5 Compaction Models

The interpretation and formulation of the different energy components, illustrated in Figure 4.14, can lead to the formation of equations regarding the constitutive behavior of materials at high-strain rates. These equations are called *compaction models,* based on the corresponding proposed consolidation mechanisms.

INTERPARTICLE MELTING MODEL

This compaction model, proposed by Schwarz et al. [30], is based on the heat dissipated because of the shock wave propagation. This energy component leads to melting a certain fraction of particles located at the boundary region. The melting energy is assumed to be provided by the total conversion of thermal energy at the particle surfaces. The shock energy is determined by the above-mentioned Equation (4.16). The energy required to heating and melting a certain fraction of material, L is given by the following equation

$$E_T = L[C_p(T_f - T_0) + \Delta H_f] \qquad (4.40)$$

where C_p is the heat capacity under constant pressure, T_f and T_0 are the fusion and ambient temperatures, respectively, and ΔH_f is the fusion enthalpy at temperature T_f.

From Equations (4.27) and (4.40) the melt fraction, L can be calculated

$$L = \frac{\frac{1}{2}PV_0\left(\frac{V_{00}}{V_0} - 1\right)}{C_p(T_f - T_0) + \Delta H_f} \tag{4.41}$$

The minimum pulse duration, t_{dmin} can also be calculated by using the above model, by considering that the minimum pulse duration should satisfy the following relationship:

$$t_{dmin} = t_s + t_c \tag{4.42}$$

where t_s is the solidification time of the melt layer and t_c the cooling time of the compact, which must be sufficient to attain certain toughness levels. According to this model, a successfully consolidated material must possess a minimum hardness, equal to half of the hardness of the corresponding solid material.

VOID COLLAPSE MODEL

This type of compaction model is based on the void collapse energy, due to the intense particle deformation. It is assumed that the powder material behaves in a plastic or viscoplastic way at high-strain rates. In general, metal and other ductile powders successfully follow these models.

Carroll and Holt proposed a model to describe the densification of porous materials, see Reference [31]. Their model is based on the notion of the collapse of a hollow sphere, loaded at the external surface by a hydrostatic pressure, P, where the parameter $\theta = \rho_0/\rho$ (the ratio between solid and current density), known as *distention,* is related to the porosity of the compact. The initial value of distention, θ_0 is expressed as a function of the sphere dimensions, a and b, see Figure 4.15(a), as

$$\theta_0 = \frac{\rho_0}{\rho_{00}} = \frac{b^3}{b^3 - a^3} \tag{4.43}$$

The viscoplastic Carroll-Holt model was modified to incorporate thermal effects, such as thermal softening, see Reference [32]. Nesterenko [33] proposed a model that simulates the void collapse process as shown in Figure 4.15(b). His model consists of a solid sphere surrounded by a concentric hollow sphere, which

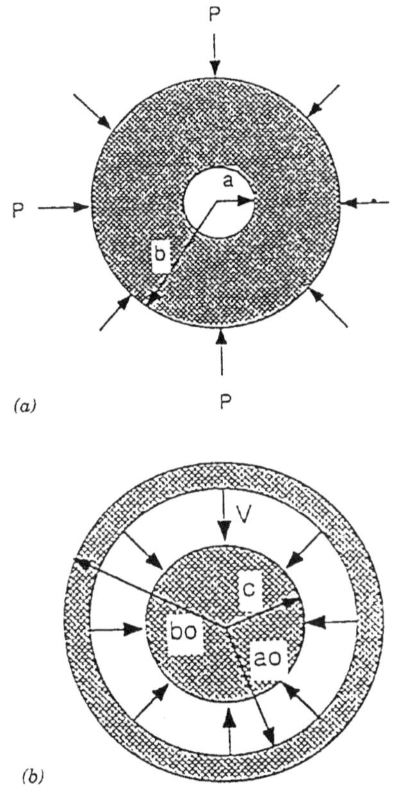

FIGURE 4.15. Void collapse models. (a) hollow sphere and (b) spherical shell including a solid sphere (from Reference [12]).

collapses on the solid sphere; the diameter of the solid sphere can be considered as the grain diameter of the compacted powder. Note that only a portion of the powder undergoing plastic deformation, i.e., the spherical shell, can be predicted. The internal solid sphere simulates the interior of the particles, which possesses negligible plastic deformation during the collapse process.

The void collapse models are based on the fact that the whole amount of the shock energy is converted into deformation and collapse energy. The void collapse energy, E_{vc}, for an idealized plastic material, may be expressed as

$$E_{vc} = \frac{2}{3}YV_0\{[\theta_0 \ln \theta_0 - (\theta_0 - 1)\ln(\theta_0 - 1)] - [\theta\ln\theta - (\theta - 1)\ln(\theta - 1)]\} \tag{4.44}$$

where Y is the yield stress of the material at elevated strain rates.

The shock pressure, P, should be greater than the value of the Vickers' Hardness, H, of the solid material to allow for the void collapse processes to be activated. For successful consolidation, the shock pressure should be greater than the value of 2 H [32].

FRAGMENTATION MODEL

Void collapse processes and particle plastic deformation are limited in the case of shock compaction of brittle ceramic powders, see Reference [4]. According to experimental observations, the dynamic loading of these type of powders, such as superconducting oxides, leads to intense fragmentation and restacking of the resulted particle fragments [2]. The various stages of this compaction mechanism can be listed as:

- fragmentation of the powder particles, due to the shock wave energy
- acceleration of the fragmented particles, resulting in restacking and pore closing
- heat dissipation, due to the frictional and impact processes, takes place at the particle boundaries; the released thermal energy leads to partial particle melting.

Note that the energy components for defect formation and chemical reaction are assumed to be zero. Thermal softening effects, leading to microscopic plasticity and shear band formation, are also neglected. A simplified approach of the above-mentioned compaction mechanism is shown in Figure 4.16(a); the "repacked" microstructure, obtained by intense grain fracturing, is shown in Figure 4.16(b).

According to this compaction mechanism, the total amount of shock energy is converted to the energy components associated with fragmentation, E_F and interface melting, E_T, i.e.,

$$E = \frac{1}{2} P(V_{00} - V) = E_T + E_F \qquad (4.45)$$

Note that the work produced because of pressure release, i.e., rarefaction, and the tensile wave reflections is assummed to be zero, i.e., $\int P \cdot dV = 0$, during relaxation. This assumption seems very reasonable in the case of brittle powder compaction.

The thermal energy component, E_T, given by the Schwarz model is described by Equation (4.40). For the calculation of the fracture component, E_F, an additional important parameter associated with the fracture process is introduced in this model, namely the *critical stress intensity factor* or *fracture toughness*, K_{Ic}, very often used in fracture mechanics. This parameter deals

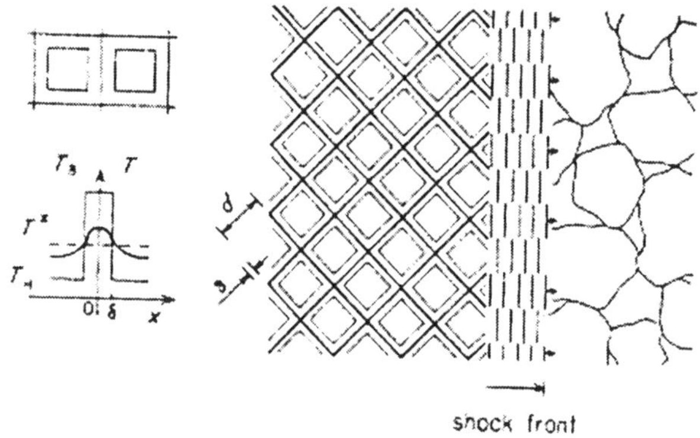

(a)

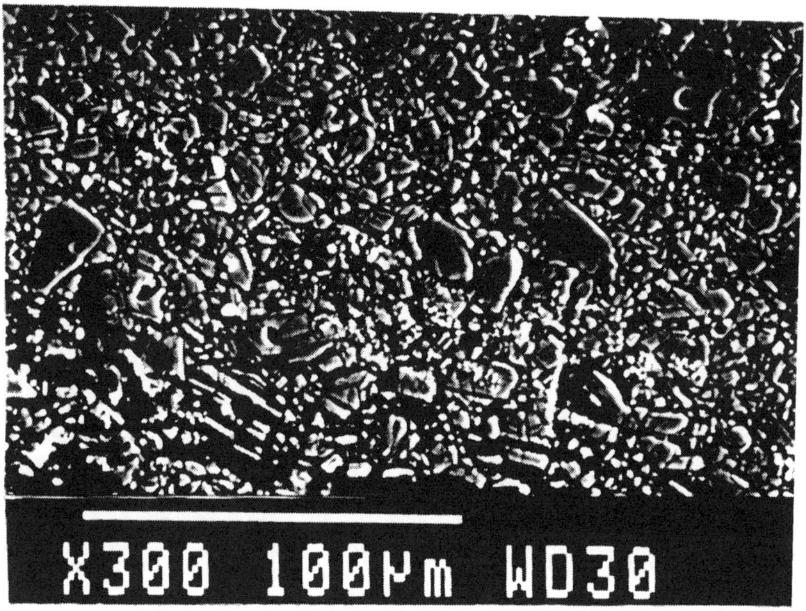

(b)

FIGURE 4.16. (a) Simple representation of fragmentation-restacking processes. (b) Scanning electron micrograph of explosively compacted YBCO ceramic.

with the fracture resistance of the material and is a measure of the critical stress (or strain energy) level for Mode I of crack propagation. Thermodynamic considerations have led to the following fracture energy relations:

$$E_F = \frac{4.2K_{Ic}^2 V_0}{\mathrm{E}}\left(\frac{1}{d_f} - \frac{1}{d_0}\right) \tag{4.46}$$

where E is the elastic modulus of the material and d_f, d_0 are the final and initial grain size, respectively. Combining Equations (4.40), (4.45) and (4.46), it yields

$$P = \frac{2}{(V_{00} - V)}\left\{\frac{4.2K_{Ic}^2 V_0}{\mathrm{E}}\left(\frac{1}{d_f} - \frac{1}{d_0}\right) + L[C_p(T_f - T_0) + \Delta H_f]\right\} \tag{4.47}$$

The accuracy shown by this above-mentioned compaction model is satisfactory compared with the experimental results. However, the lack of thermochemical data renders difficulties for the calculation of the thermal energy component, for relatively medium shock pressures, i.e., not exceeding 10 GPa.

4.4.6 Shock-Induced Failures

CRACKS

Dynamic (shock wave) compaction encounters, very often, for macro- and microdefects, are classified below. In the case of the cylindrical axisymmetric setup, the convergence of shock waves at the central axis may be responsible for five types of cracks, see Figure 4.17 and Reference [13], namely:

- *Helicoidal cracking:* This type of cracking is due to the instability of compressive shear stresses and is prone to low ductility of the compacted materials.
- *Mach-stem formation:* This is due to the compressive stress fields developed as a result of overcompaction. In the middle of the tube, the generated pressure is too high, resulting in material turbulence, with subsequent melting or even blown-out material ejection.
- *Circumferential cracking:* This is due to the tensile stress field developed, when reflected and transmitted shock waves are superimposed, resulting, therefore, in circumferential cracking. The tensile wave, reflected at the metal interface of the sheathed component, is, sometimes, strong enough to cause circumferential failures, the so-called *spalling fractures.*
- *Radial cracking:* This is due to the tensile stress field developed, just before the pressure release, when the material near the metal/ceramic interfaces is under higher compressive stress than the powder material in the center of the tubular component. As a result, radial cracks are formed initiating at the center of the tube.

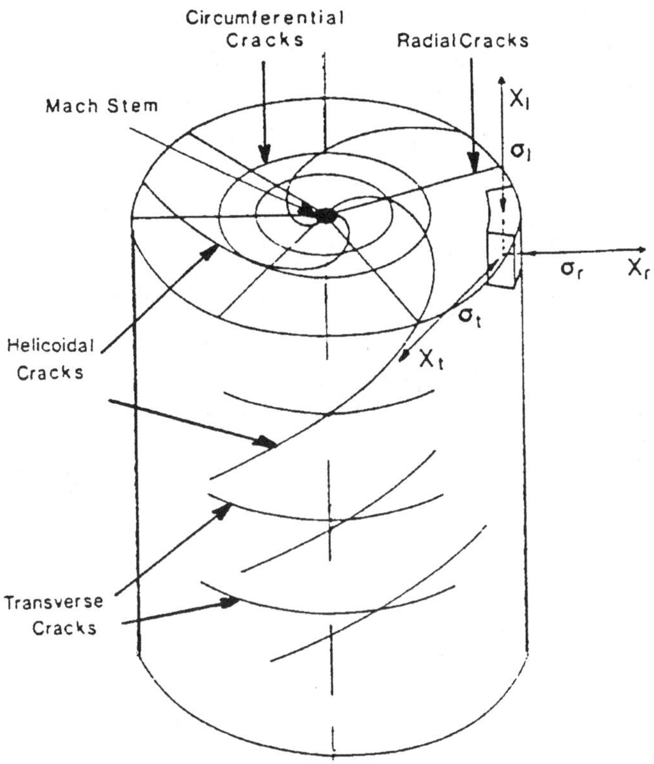

FIGURE 4.17. Possible macrocracks generated after explosive compaction (from Reference [13]).

- *Transverse cracking:* This is due to the tensile stress resulting from:
  —the impedance mismatch between the compacted material and the bottom plug,
  —the stretching of the cylinder, and
  —the thermal stresses during cooling.

Figure 4.17 illustrates all the basic cracks induced by dynamic compaction method, whereas in Figure 4.18 presented are some characteristics fracture patterns of explosively compacted YBCO powders; see below.

DISLOCATIONS

Shock wave propagation in powder materials encounters the development of characteristic microscale defects, such as several types of dislocations. The detection of those dislocation networks can be achieved by means of high-resolution electron microscopy (HREM/TEM), see References [34,35].

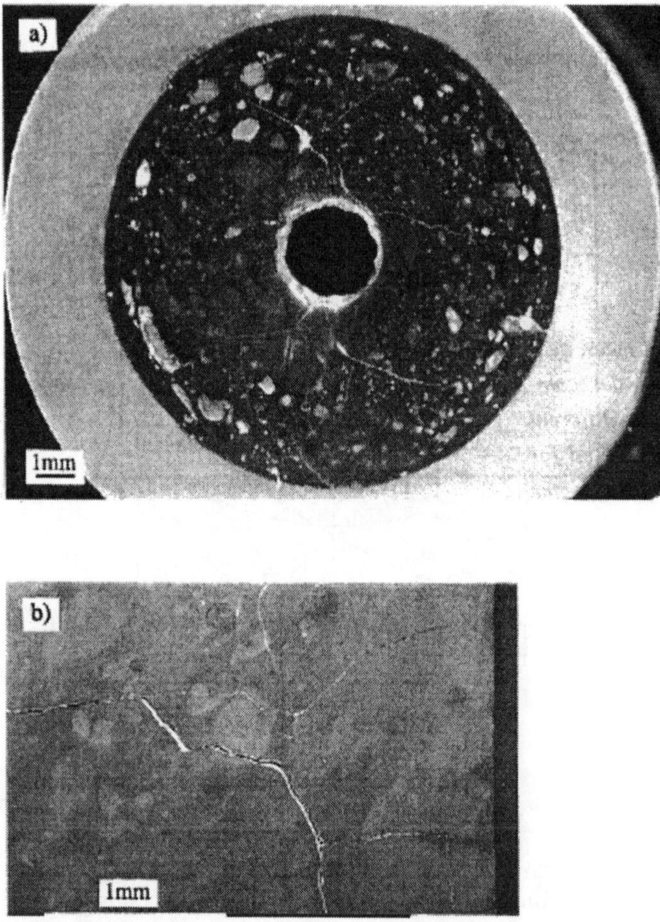

FIGURE 4.18. Optical micrographs showing (a) the Mach stem hole formation and (b) the presence of helicoidal cracks (from References [34] and [35]).

The basic types of dislocations, observed in explosively compacted YBCO samples, can be listed as:

- *Type A dislocations:* This type of dislocation has been also observed in melt-textured and cold-pressed YBCO samples; they are characterized by the [100] or [010] Burgers vectors, gliding on the crystallographic planes (001), see Figure 4.19.
- *Type B dislocations:* They have been mainly observed in explosively YBCO samples and are characterized by the formula [110](110). The formation of this type of dislocations, with Burgers vectors [110] and [010], is a result of dislocation reaction, see Figure 4.20.

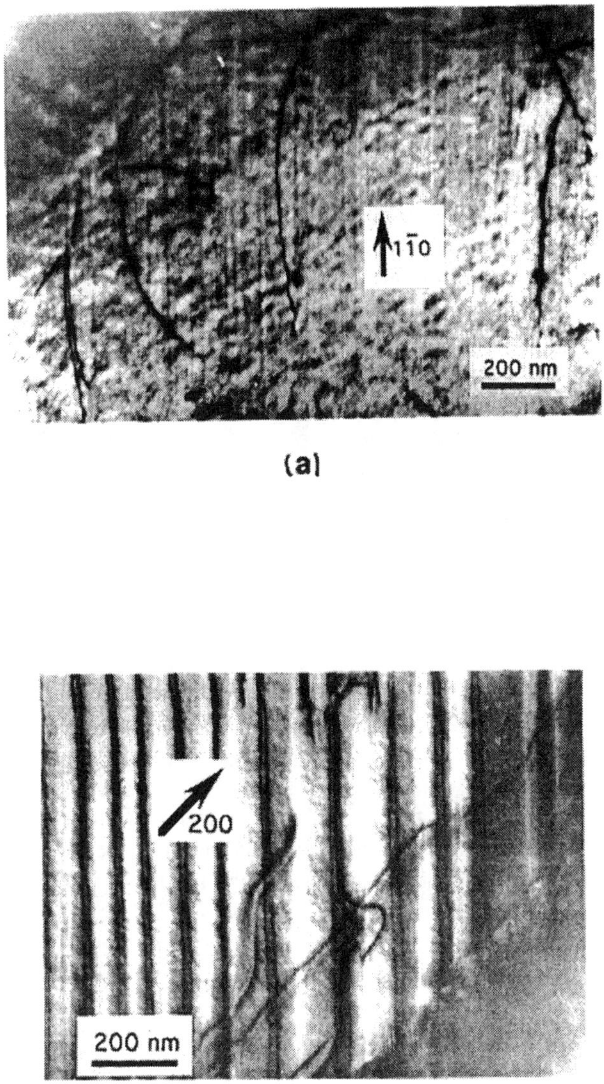

(a)

(b)

FIGURE 4.19. (a) (110) TEM dark-field image of the initial powder: type A dislocations. (b) (200) TEM dark-field image in explosively compacted YBCO (R = 0.78): type A screw dislocations (from References [34] and [35]).

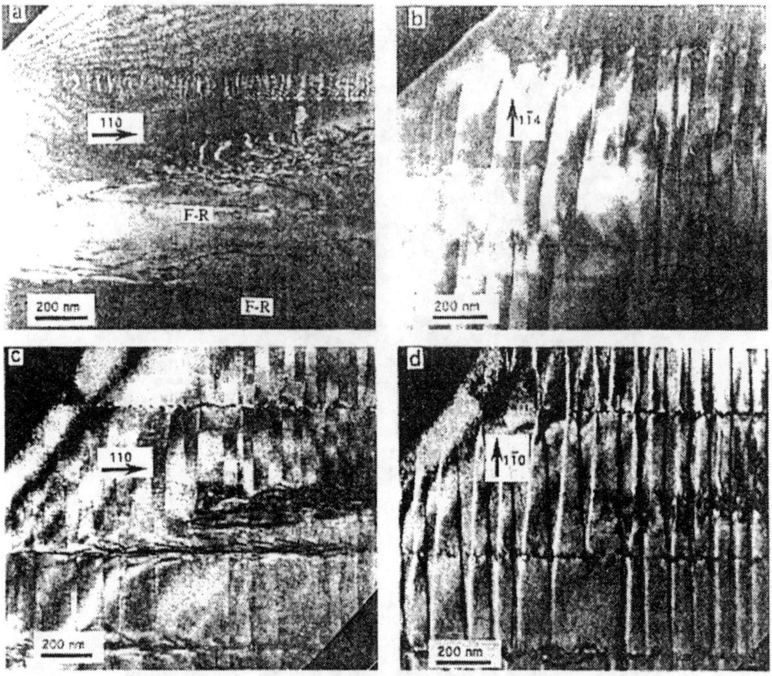

FIGURE 4.20. Diffraction images of type B dislocations in explosively compacted YBCO (R = 0.78) (from References [34] and [35]).

- *Type C dislocations:* This type of dislocation, observed in explosively compacted YBCO samples, is formed by gliding of dislocations [100] and [010] on the crystallographic planes (010) and (100), respectively, see Figure 4.21; these dislocations are limited on the (001) plane ($\perp$ c axis).

4.4.7 Calculation of Explosive Compaction Parameters

COMPACTION/CLADDING OF METAL/HTS/METAL SANDWICH PLATES

Consider the experimental configuration for the cladding/compaction of metal/superconductive ceramics shown in Figure 4.22. The cladding/compaction parameters are determined for the metal/HTS/metal sandwich plate due to fabrication and can be determined as follows:

(1) The velocity at impact, v_p of the flyer plate, when accelerated after detonation and collided on the parent plate can be estimated by using the Gur-

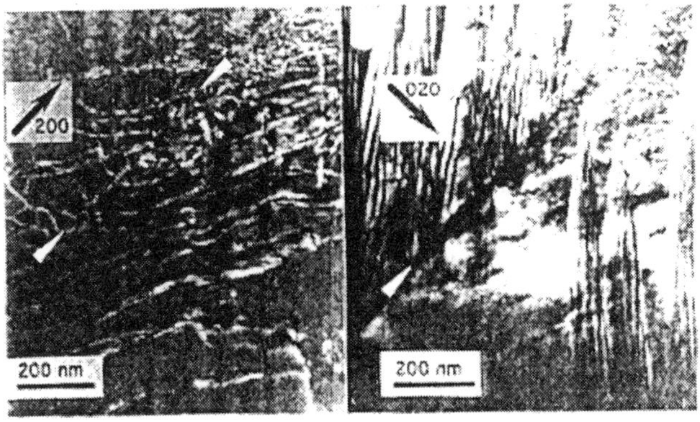

FIGURE 4.21. Type C dislocations in explosively compacted YBCO (R = 1.39) (from References [34] and [35]).

ney model, see Reference [35] for granular type explosives, such as ammonium nitrate. Following the Notation section,

$$v_p = 1.2v_d \frac{(1 + 1.2\zeta)^{1/2} - 1}{(1 + 1.2\zeta)^{1/2} + 1} \tag{4.48}$$

where $\zeta = (\rho_e h_e)/(\rho_f h_f)$, ρ_f is the density of the flyer plate, h_e the thickness of the explosive layer and h_f the thickness of the flyer plate.

(2) The collision angle can be estimated from the governing equation of the explosive cladding process

$$f \cdot v_p = 2v_d \sin\left(\frac{\beta}{2}\right) \tag{4.49}$$

The constant f, related to the standoff distance, is taken equal to unity, when the standoff distance is sufficient for the velocity of the cladding plate at impact, or estimated from Equation (4.49), to be attained. For standoff distances smaller than the critical value, $f < 1$.

(3) In the case of explosive cladding, when collision takes place between the flyer plate and the parent plate, the developed peak pressure at impact, P can be estimated from the Rankine-Hugoniot relation:

$$P - P_0 = \rho_f u_s u_p \tag{4.50}$$

When the flyer and parent plates consist of the same material and the ambient pressure, P_0 is too small, compared with the shock pressure, P, Equation (4.50) leads to the following equation:

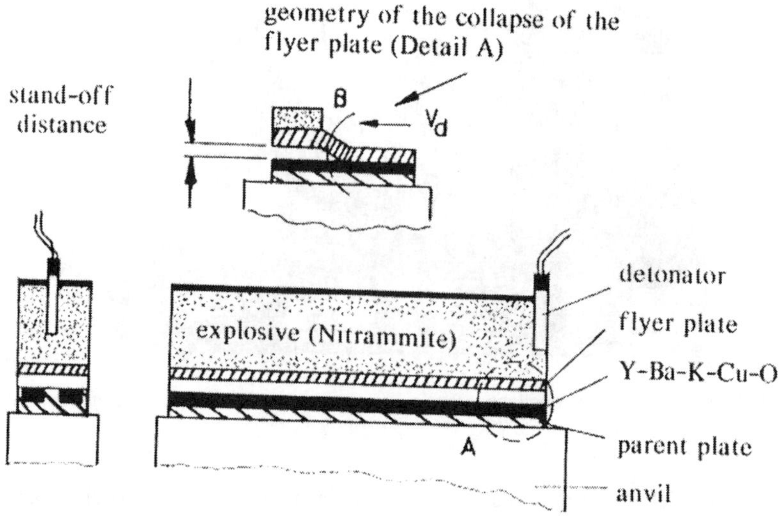

(a)

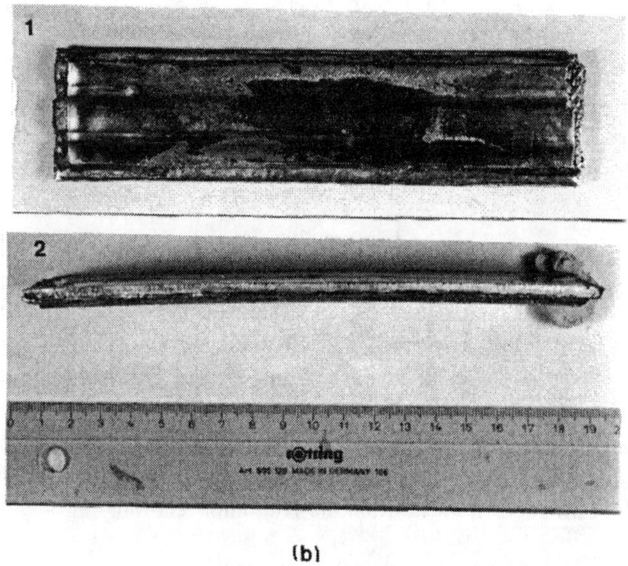

(b)

FIGURE 4.22. (a) Experimental setup of the plane compaction/cladding configuration. (b) 1: Top view and 2: side view of the Ag/YBKCO/Ag plate.

$$P = \frac{1}{2} f \rho_f u_s v_p \tag{4.51}$$

For the cladding geometry with an intermediate superconducting layer, shown in Figure 4.22, the shock pressure, developed between the flyer cladding plate and the superconducting layer of the parent plate, can be estimated from a proposed modified Rankine-Hugoniot relation, taking into account the related impedance ratio $Z' = Z_f/Z_{sc}$

$$P = \frac{\rho_f u_s f v_p}{1 + Z'} \tag{4.52}$$

where $Z_f = \rho_f u_s$ and $Z_{sc} = \rho_{sc} u_{sc}$ is the impedance of the flyer metallic plate and of the superconductive ceramic portion of the parent plate, respectively; ρ_{sc} is a mean value of the density of the superconducting ceramic during explosive cladding, whereas u_s, u_{sc} can be approximated to sound velocities of the elastic stress waves, propagated in the metal and in the superconducting material, respectively.

(4) The kinetic energy per unit area of the flyer plate may be estimated from

$$E_k = \frac{m_f (f v_p)^2}{2A_f} \tag{4.53}$$

while the released energy from the shock wave compaction, stored in the superconducting material as additional internal energy, can be estimated from

$$E = \left(\frac{P + P_0}{2}\right)\left(\frac{\rho - \rho_{00}}{\rho \cdot \rho_{00}}\right) \tag{4.54}$$

where ρ_{00} and ρ are the initial and the final density of the superconductive material, respectively.

AXISYMMETRIC EXPLOSIVE COMPACTION OF METAL/HTS/METAL BILLETS

For the axisymmetric explosive powder compaction, in a cylindrical setup, see Figure 4.2, the velocity, v_p at impact, when the shock wave, resurting from the explosion, hits the outer periphery of the metal tube propagating into the porous medium, can be estimated by using the Gurney model [35], as

$$v_p = \left[\frac{6E}{5R + \frac{2}{r_c} \cdot \frac{1}{R} \cdot (2r_e + r_c) + \frac{2r_c}{r_e + r_c}} \right]^{1/2} \tag{4.55}$$

where r_e is the outer radius of the explosive compaction setup, r_c the radius of the ceramic core before compaction, E the chemical energy of the explosive per unit mass and R the impact mass ratio designated as the ratio explosive mass/metal container mass.

4.5 EXPLOSIVE POWDER COMPACTION OF HIGH-T_C CERAMICS

The experimental details pertaining to the explosive compaction of superconducting powders of the Y-Ba-Cu-O and Bi-Sr-Ca-Cu-O systems are given below for both the 3D and the axisymmetric configuration techniques. The main features of the explosively induced microstructures, phase transformations and stoichiometric changes are reviewed. Metal/ceramic interface phenomena and localized plastic deformation and/or fracture are also reported and discussed in terms of shock wave propagation and interaction in solid/porous media.

4.5.1 Preparation of Powders Prior to Compaction

Y-Ba-Cu-O

For the preparation of the YBCO-type ceramic powders, the solid-state reaction technique, e.g., the ceramic method was used. The powder mixture, with the nominal composition of $YBa_{2-x}K_xCu_3O_y$ (x = 0–0.1), was prepared by using stoichiometric quantities of Y_2O_3, $Ba(OH)_2 \cdot 8H_2O$, $KF \cdot 2H_2O$, CuO and $AgNO_3$. The mixture of these materials was ground with a pestle in an agate mortar and homogenized by adding alcohol. Copper oxide (CuO) was obtained by the calcination of $Cu(OH)_2 \cdot CuCO_3 \cdot nH_2O$.

After drying the homogenized powder mixture, pellets of 25 mm in diameter and 3 mm in length were formed by pressing with a hydraulic press; the pressure range applied was 25–51 MPa. The pellets were sintered for 3–10 h with intermediate grindings in a programmable electric furnace in flowing oxygen (15–50 dm^3/h) at a temperature of 950°C followed by slow cooling in the furnace. Subsequently, sintered pellets were pulverized in an agate mortar; the final grain size varied between 63 and 900 μm.

The Ag content, originated from $AgNO_3$, was about 10% w/w in the final powder mixture. Silver inclusions, homogeneously distributed in the material, serve to the improvement of ductility, the minimization of porosity and the increase of fracture toughness and thermal fatigue strength [37]. The Ag-HTS materials are expected to possess also enhanced formability. In general, no chemical interaction between the silver and the ceramic powder occurred in the present studies.

The beneficial effects of alkali and fluorine doping regarding the superconducting properties of the YBCO material have been examined in the previous chapter (Section 3.4). A scanning electron micrograph of the starting alkali doped ceramic powder is shown in Figure 4.23, whereas the corresponding X-ray diffraction pattern, revealing the presence of the 123 orthorhombic superconducting phase, is presented in Figure 4.24. The magnetic susceptibility measurements show diamagnetic transitions at onset T_c ranging from 90 to 102 K. The K-doped YBCO powder was used for both plane and axisymmetric explosive compaction experiments.

Bi-Sr-Ca-Cu-O

The powder used was prepared from the appropriate mixture of Bi_2O_3, $SrCO_3$, $CaCO_3$, CuO and PbO to provide the nominal composition of $Bi_{1.5}Pb_{0.5}Sr_2Ca_2Cu_3O_x$. Lead is added to improve the reaction kinetics of the formation of the 2223 phase. The powders were mixed in an agate mortar and calcined at 880°C for 3 h in air. The resultant compound was pulverized. Paste, prepared by thoroughly mixing the powder with organic material, was screen printed on a silver (Ag 999) substrate with dimensions of 200 × 40 mm and 2-mm thickness. After drying the thick superconducting film, the organic material was removed by heating the thick film at 500°C for 2 h. The film thickness was in the range of 100–250 μm, and its width varied between 20 and 26 mm. The same technique of preparing thick superconducting film was previously used, see Reference [38].

The microstructure of the BPSCCO ceramic film is shown in Figure 4.25 (top view). The grain size varied between 0.2 and 10 μm with an average value of 2 μm. XRD analysis revealed the presence of various phases, see Figure 4.26(a). The percentages (% w/w) of each phase were calculated according to the Rietveld analysis from XRD data, as 2223 phase: 78%; 2212 phase: 12%; 2201 phase: 9%; Ca_2PbO_4: 1%, see also Reference [39].

The ac-susceptibility measurements indicated two relatively sharp diamagnetic transitions at T_{c1} = 108 K (ΔT_{c1} = 12 K) and T_{c2} = 86 K (ΔT_{c2} = 8 K) that correspond to the high-T_c 2223 and low-T_c 2212 phase, respectively, see Figure 4.26(b). The ratio of the diamagnetic transition heights represents the volume fraction between these phases [39]. This preparation technique is used as the first stage before explosive compaction, when the 3D configuration setup is used.

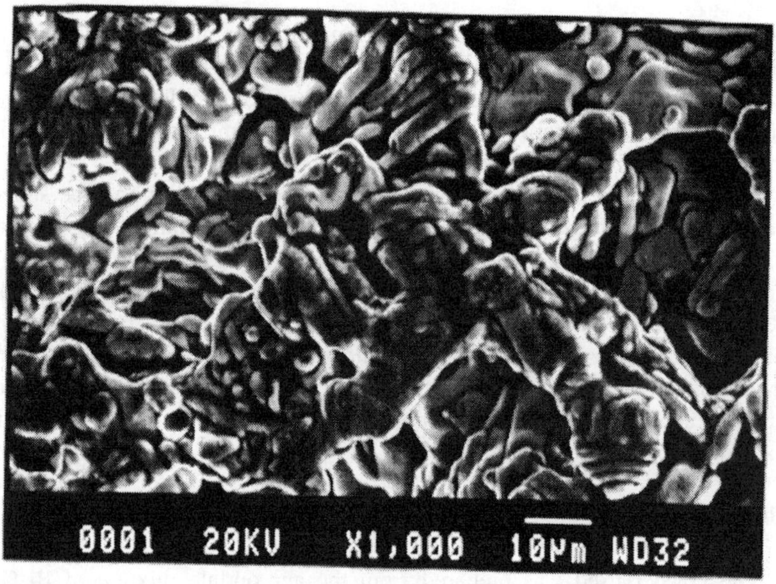

(a)

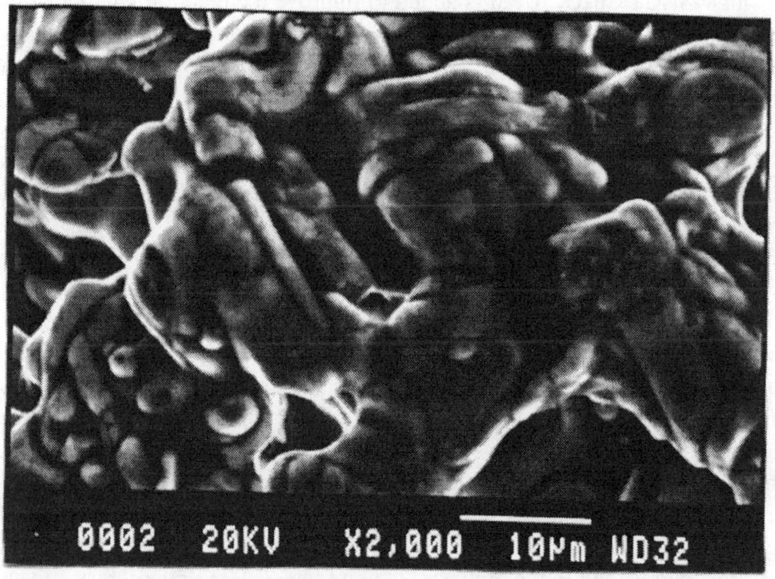

(b)

FIGURE 4.23. Scanning electron micrographs showing (a) the microstructure of the YBCO

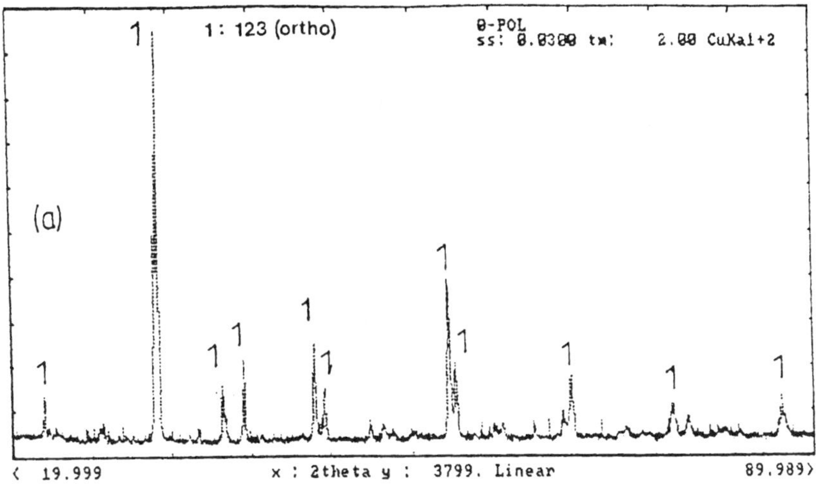

FIGURE 4.24. XRD pattern of YBCO sintered ceramic.

4.5.2 Compaction/Cladding of Metal/YBCO Plates

Explosive compaction/cladding was used for the fabrication of the superconductor/silver composite plates for producing strips by subsequent cold-rolling [2]. The parallel cladding configuration, with an initial setup angle of 0° and a standoff distance of 2 mm, was used, as shown schematically in Figure 4.22. The explosive used was the "Nitrammite" high explosive, in granular form, consisting of ammonium nitrate in a percentage above 98%, with 1.0 g/cm^3 bulk density and 4000 m/s detonation velocity. An explosive mass of 200 g was used, forming a layer of 35-mm thickness above the flyer plate, as shown in Figure 4.22(a), and providing an impact mass ratio, R of 1.2. The dimensions of the flyer and the parent silver plates were 190 × 30 × 2 and 200 × 40 × 5 mm, respectively. Two identical longitudinal grooves of dimensions 200 × 5 × 2.5 mm were machined in the parent plate. To form a sound component, magnesium oxide (MgO) powder was placed at the ends of each groove, before compaction, avoiding, therefore, fracturing of the superconducting ceramic near the edges, due to the superposition of the reflected shock waves; see Section 4.4.6 above. The rest of the groove was filled by superconducting powder.

Three types of powders are used:

- 1st type (4.6 g, ρ_{00} = 39%): high-T_c ceramic powder synthesized without silver nitrate addition;
- 2nd type (3.4 g, ρ_{00} = 57%): high-T_c ceramic powder prepared by adding silver nitrate in the reacting mixture;

FIGURE 4.25. Scanning electron micrographs showing (a) the microstructure of the BSCCO thick film and (b) detail of (a) (top view).

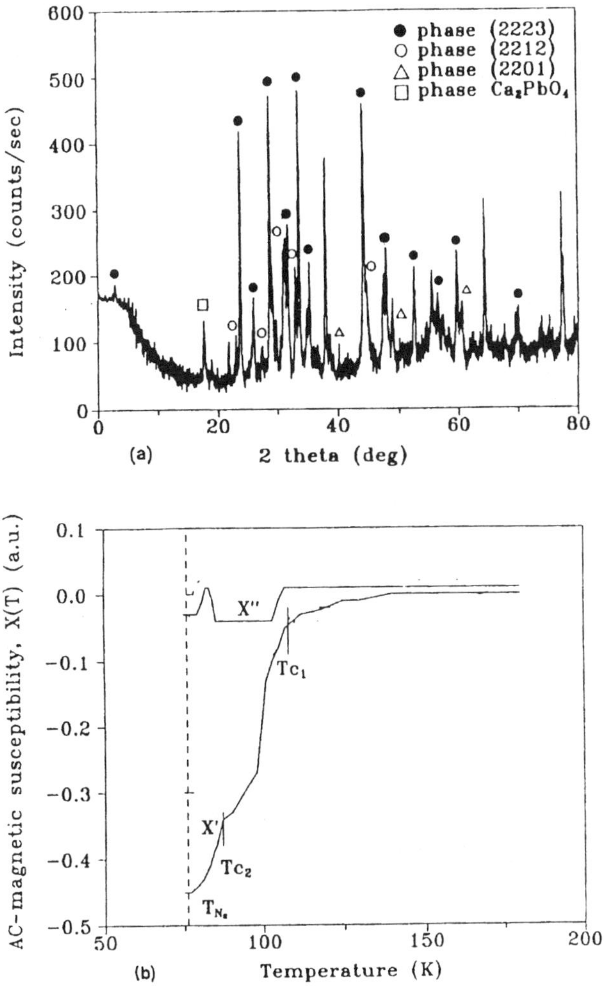

FIGURE 4.26. (a) XRD pattern of the BSCCO thick film. (b) The ac-susceptibility curves of the BSCCO initial powder.

- 3rd type (1.4 g, ρ_{00} = 27%): stoichiometric mixture of nominal composition $Y(Ba,K)_2Cu_3O_y$ + 10% w/w $AgNO_3$.

The theoretical density, ρ_0 was considered to be around 6.28 g/cm^3, whereas the bulk density, after the explosive compaction, was about 85% of the nominal density.

Explosions were conducted in sand, by firing an electric detonator, resulting in the collapse of the flyer plate and the consolidation of the powder. The explosive cladding parameters for the 2nd type of powder ($f = 0.8$, $u_s = 3650$ m/s, $\rho_{sc} = 4.45$ g/cm^3, $u_{sc} = 4000$ m/s) have been calculated by using Equation (4.54) and are tabulated in Table 4.1.

Longitudinal and cross sections were cut off from the explosively cladded plates and subjected to a standard metallographic preparation, see Figure 4.22(b). The microstructural examination was performed by optical and scanning electron microscopy (SEM).

Macro- and optical microscopic observations of the explosively cladded plates revealed several macro- and microdefects. The microstructure of the three types of explosively consolidated powder is shown in the optical micrographs of Figures 4.27, 4.28 and 4.29, respectively. The most dense and sound microstructure was observed for the 2nd type YBCO powder. This type of compacted powder is also characterized by the finest grain size, compared with the other two types of powders.

Micropores and intense cracking were observed for the 1st type of compacted powder, mainly developed because of the shock wave passage through the powder body, see Figure 4.30(a)–(c); these cracks may be attributed to the peak shock pressures developed during explosive compaction, combined with the brittle response of the ceramic material; compare also with the SEM micrograph of the corresponding consolidated powder presented in Figure 4.23. The higher porosity appeared in the 3rd type compacted powder; macropores and an almost complete destruction of the encapsulated compact are indicated in Figure 4.30(d).

The soundest macrostructure, showing sufficient mechanical integrity, is related to the 2nd type of the explosively compacted powder. Optical microscopy revealed intense fragmentation of the initial grains, indicating, therefore, sufficient powder compaction; see Figure 4.31(a). The very good adhesion of the

TABLE 4.1. Explosive Compaction/Cladding Parameters (YBCO).

Parameter	Value
Dimension of flyer plate (Ag)	190 × 30 × 2 (mm)
Dimension of parent plate (Ag)	200 × 40 × 5 (mm)
Type of explosive	Nitrammite
Detonation velocity	4000 (m/s)
Initial setup angle	(0) (°)
Collision angle	14 (°)
Standoff distance	2 (mm)
Thickness of explosive	35 (mm)
Impact mass ratio	1.2
Velocity of flyer plate	1286 (m/s)
Peak shock pressure in Ag/Ag	20 (GPa)
Peak shock pressure in HTS (2)	13 (GPa)
Kinetic energy of flyer plate	1.1×10^7 (J/m^2)
Increase of internal energy HTS(2)	5.1×10^5 (J/kg)

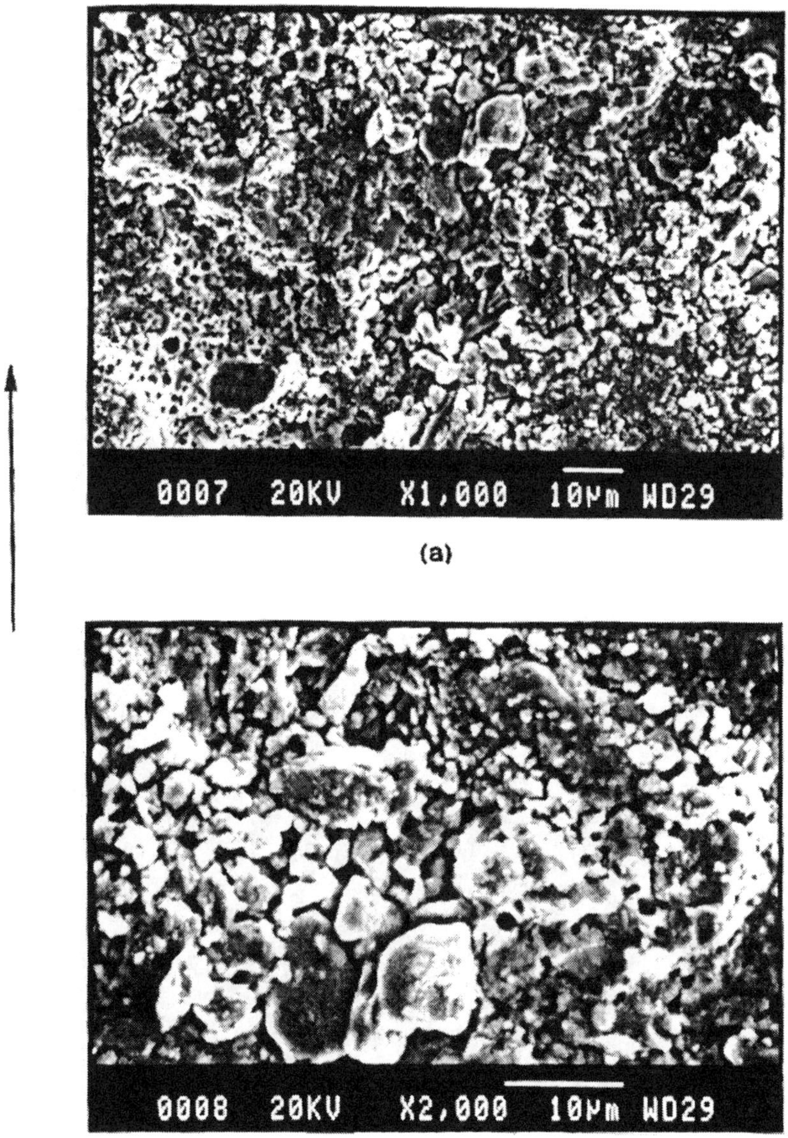

FIGURE 4.27. Scanning electron micrographs showing (a) the microstructure of the 1st type YBCO compacted powder and (b) detail of (a). The arrow indicates the detonation direction.

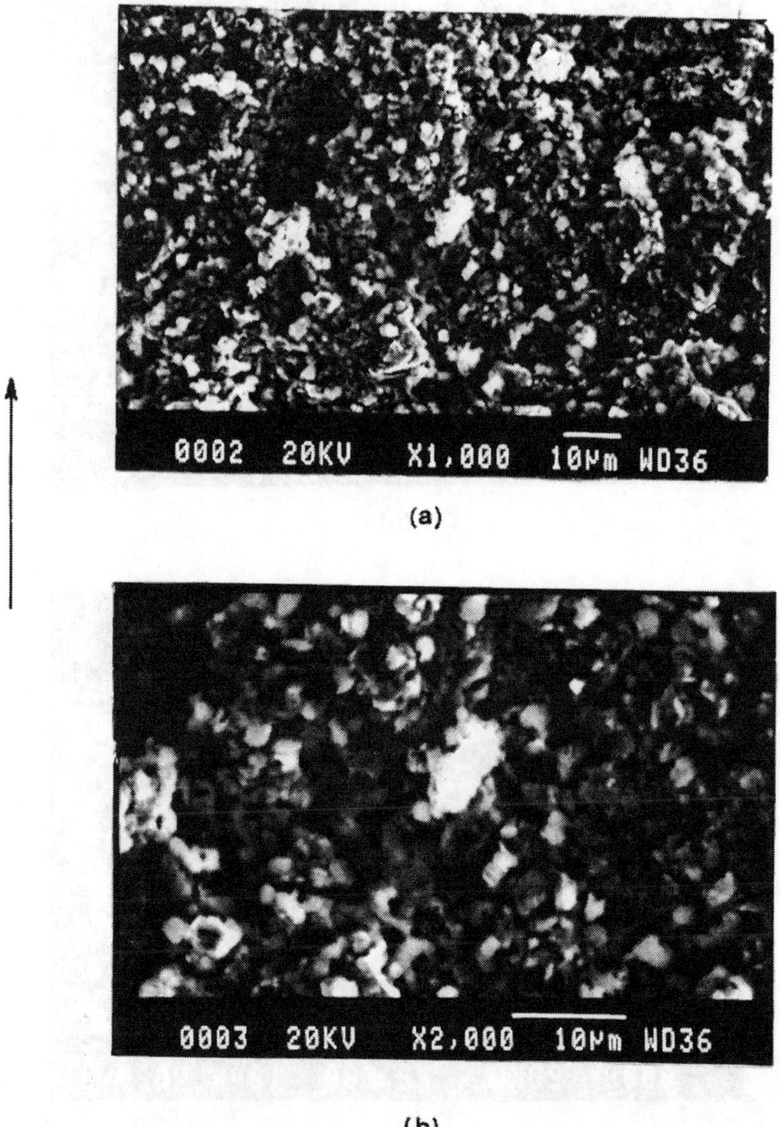

FIGURE 4.28. Scanning electron micrographs showing (a) the microstructure of the 2nd type YBCO compacted powder and (b) detail of (a). The arrow indicates the detonation direction.

(b)

FIGURE 4.29. Scanning electron micrographs showing (a) the microstructure of the 3rd type YBCO compacted powder and (b) detail of (a). The arrow indicates the detonation direction.

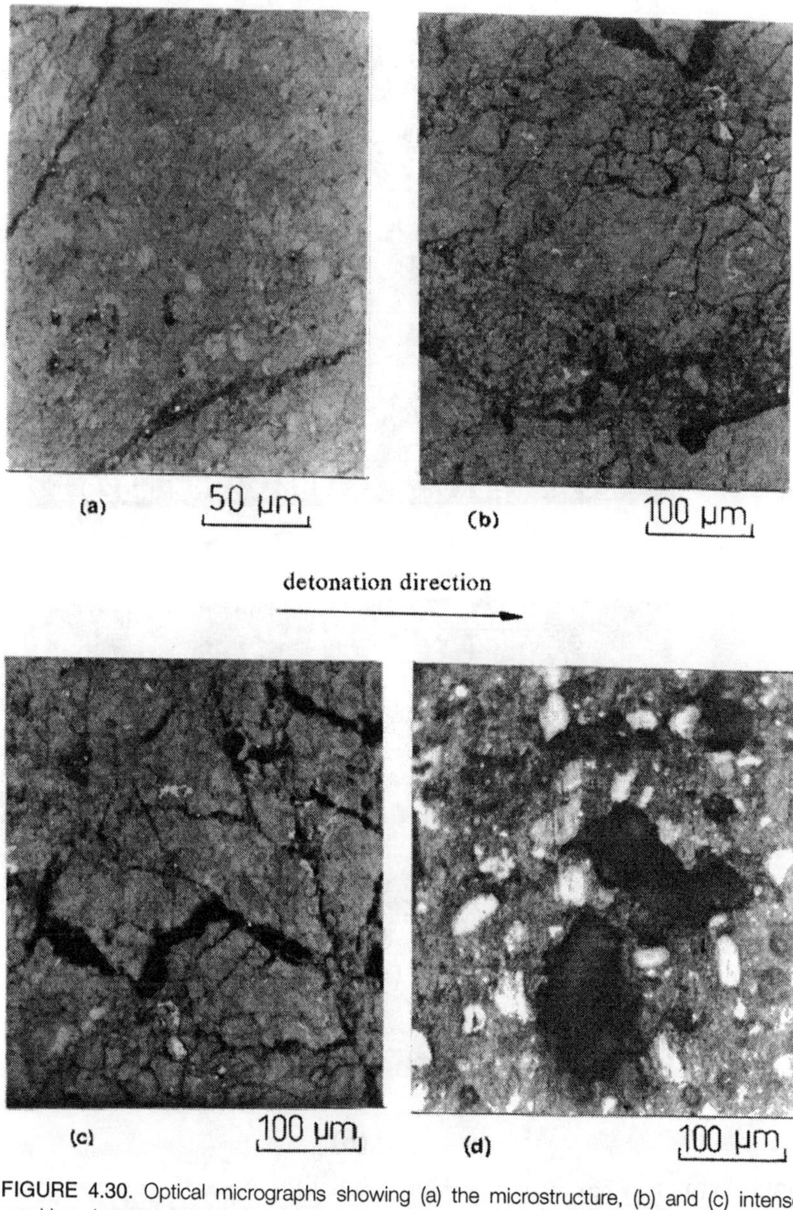

FIGURE 4.30. Optical micrographs showing (a) the microstructure, (b) and (c) intense cracking phenomena of the 1st type compacted powder and (d) the microstructure of the 3rd type compacted powder.

ceramic to both the flyer and the parent plates is confirmed in Figure 4.31(b). Detailed metal/ceramic interface features are also shown in Figure 4.32.

Shear cracks were only locally found in the powder body, indicating the occurrence of spalling failures, e.g., fractures mainly attributed to the reflection of strong tensile stress waves at the ceramic/metal interfaces, which may lead to further grain fragmentation and cracking at about the maximum shear stress direction, i.e., 45° to the longitudinal direction, see Figure 4.32(b). However, cracking was less intense in the case of the 2nd type ceramic powder, because of silver content, which, in a percentage more than 10%, may significantly influence the soundness of the fabricated ceramic; it improves the static and dynamic mechanical properties of the material, such as ductility, fracture strength and fracture toughness and thermal fatigue strength, and it also contributes to the minimization of the porosity without degrading the superconducting properties, T_c, J_c. Scanning electron micrograph of the 2nd type of the explosively consolidated powder, presented in Figure 4.28, indicates the fine-grained microstructure that resulted from the shock wave passage through the porous material.

Microstructural inhomogeneities, observed by scanning electron microscopy near the longitudinal edges of the superconducting channel, are shown in Figure 4.33(a) and (b); they are needle-like crystallites, with chemical composition, found by using energy-dispersive spectrometry (EDS), corresponding to the barium-rich phase, which might be barium oxide (BaO) or barium cuprate ($BaCuO_2$); see Figure 4.33(c). The possible barium phase crystallization and stabilization may be attributed, in this case, to the presence of silver and to the elevated shock pressure (around 13 GPa) and temperature, developed during the shock wave passage through the powder material.

The peak shock pressure significantly affects the soundness and the superconducting properties of the fabricated component. For relatively small peak shock pressures, i.e., less than 3 GPa, an inadequate compaction is achieved, caused by small-scale particle melting and subsequent welding, with a slight reduction of porosity, whereas peak shock pressures greater than 20 GPa result in excessive melting between the grains, reducing the porosity, but causing amorphous welded areas due to the rapid solidification after compaction. Moreover, many lattice defects, phase transformations and oxygen loss are enhanced, because of the passage of the shock wave front through the porous material, resulting, therefore, in the degradation of the residual superconductivity, see References [14,17]. Such waves may also cause spalling at the interface, due to their reflection into strong tensile waves.

X-ray diffraction patterns, obtained by applying the X-ray diffraction technique (CuKα radiation) on the superconducting pellet, revealed the following, see Figure 4.34(a) and (b):

- The diffraction peaks of the explosively compacted ceramics were weaker and broader, compared with the initial powder, indicating, therefore, the

FIGURE 4.31. Optical micrographs showing (a) the microstructure and (b) a shear crack initiated at the interface Ag/YBCO (2nd type compacted powder).

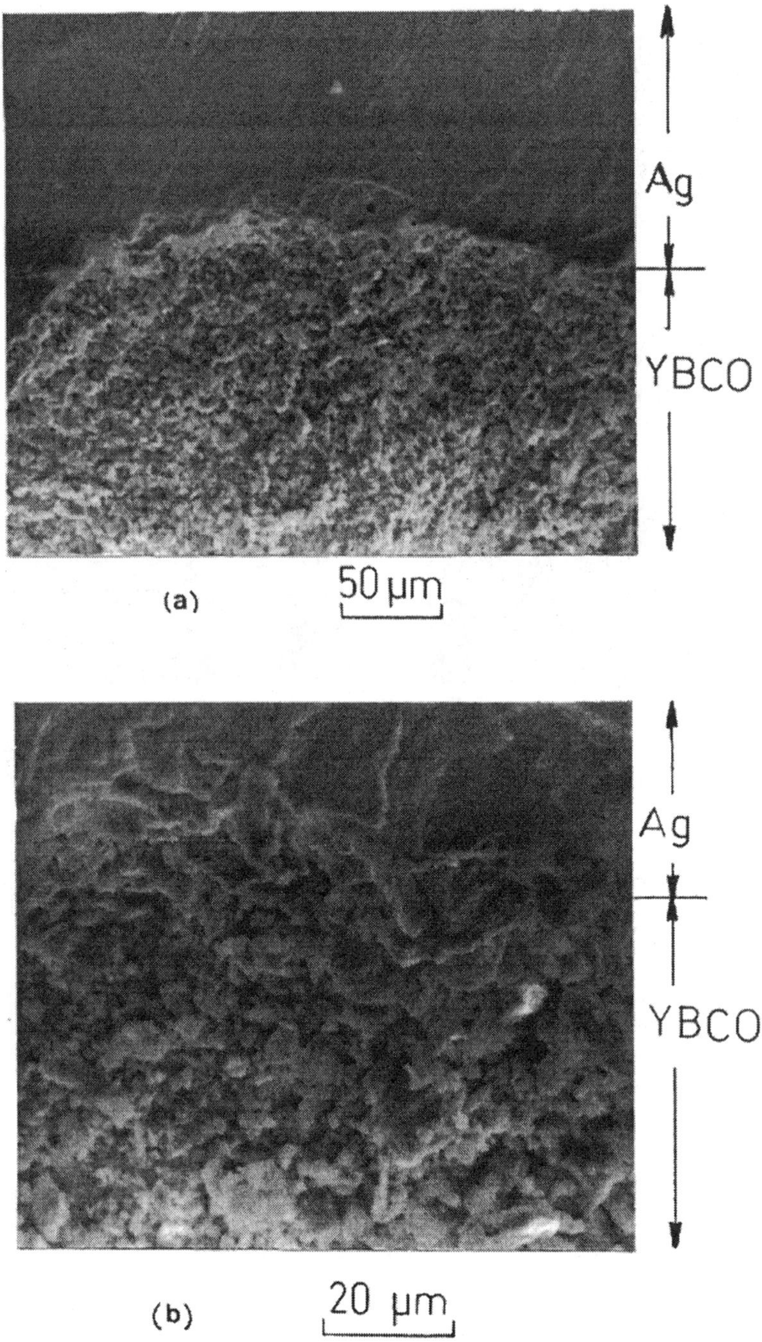

FIGURE 4.32. Scanning electron micrographs showing (a) the Ag/YBCO interface and (b) a detail of (a) (2nd type compacted powder).

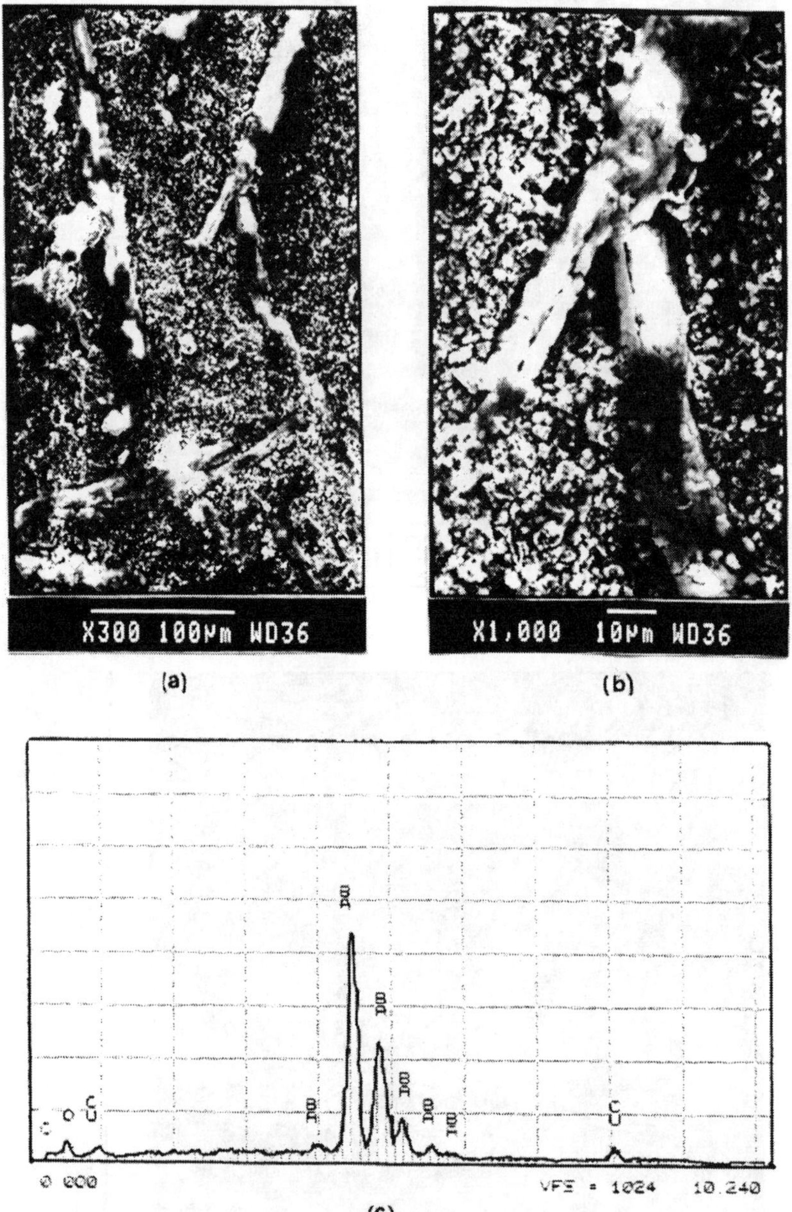

FIGURE 4.33. (a) Scanning electron micrograph of needle-like crystallites in YBCO compacted powder (2nd type compacted powder). (b) Detail of (a). (c) EDS spectrum of these crystals.

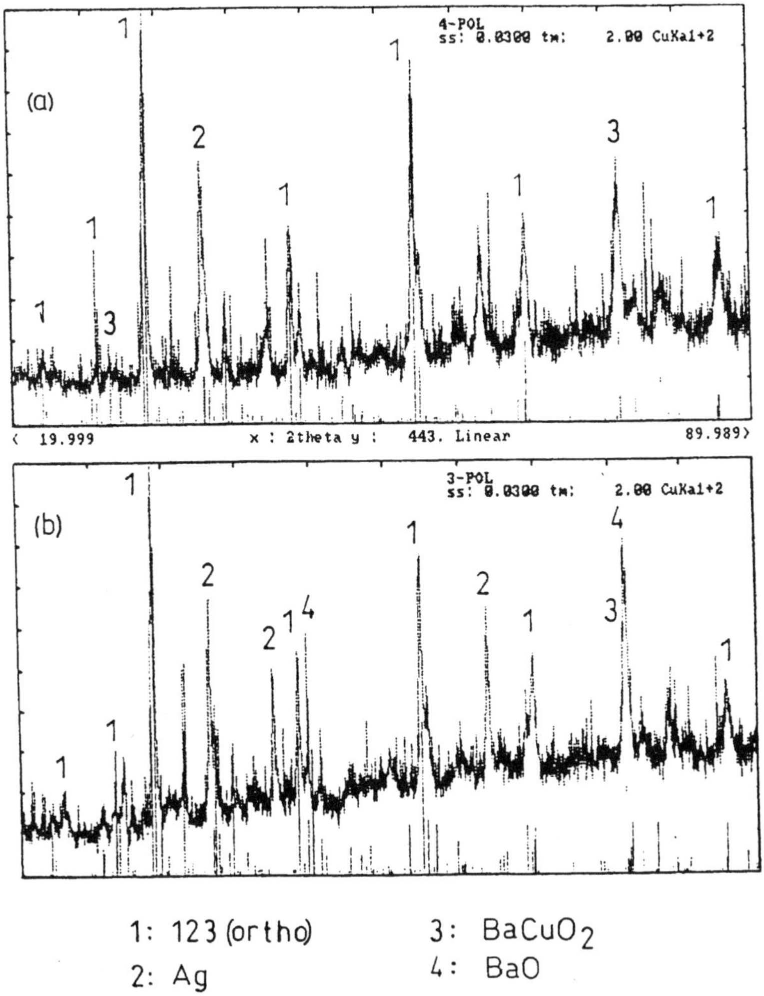

FIGURE 4.34. XRD patterns of (a) 1st type and (b) 2nd type compacted powder.

presence of shock-induced crystal defects and finer microstructure; the presence of grain boundaries widens significantly the diffraction peaks.

- Partial transformation of the orthorhombic to tetragonal phase were not present in all powder types examined, as indicated from the absence of the characteristic tetragonal phase reflection peaks.
- The XRD patterns of the 1st type of the explosively compacted powder, see Figure 4.34(a), indicated the presence of the orthorhombic superconducting phase, combined with a secondary phase, namely barium cuprate

($BaCuO_2$), whereas the related XRD pattern of the 2nd type of the explosively compacted powder, see Figure 4.34(b), revealed the presence of the orthorhombic superconducting phase, combined with silver and secondary phases, such as barium oxide and barium cuprate.

- No superconducting phase was revealed in the case of the 3rd type compacted powder; only the initial mixture components are apparent.

Thermal analysis was performed by using the standard thermogravimetric technique (TGA) using a thermogravimetric analyzer. The oxygen percentage dealt with the total weight loss that resulted from the thermal decomposition of the superconducting phase during heating in argon atmosphere from 50° to 920°C, with an increasing temperature rate equal to 4°C/min.

Thermogravimetric analysis was applied on the 1st and 2nd types of the explosively compacted powder and on the initial free of silver powder for comparison purposes. A TGA diagram, showing the evolution of weight loss as a function of temperature for the three kinds of powder examined, is presented in Figure 4.35. The total weight loss for the initial powder was about 3%, corresponding to the additional oxygen release and leading, therefore, to the formation of the stable tetragonal nonsuperconducting phase, namely the $YBa_2Cu_3O_6$. The weight loss for the explosively fabricated materials was nearly the same, around 5.5%, see Figure 4.35, indicating the release of oxygen not only from the orthorhombic superconducting phase but also from the secondary phases, such as the barium cuprate. The nonsymmetric shapes and the many inflection points of the corresponding TGA curves indicate the possible decomposition of the secondary phases, increasing, therefore, the oxygen loss.

The superconducting behavior of the explosively fabricated shapes was evaluated by using the dc-magnetization technique, based on the diamagnetic properties of superconductors, i.e., the Meissner effect. The magnetization measurements result in the construction of two basic curves, see Figure 4.36(a). The first curve is the zero-field-cooled (ZFC) curve, comprised of the magnetization measurements taken by heating the sample in a constant magnetic field (200 Oe) after zero field cooling; the second curve is the field-cooled (FC) curve, consisting of the magnetization measurements obtained by cooling the sample in a constant magnetic field (200 Oe). The transition temperature, T_c, of the 1st and 2nd types of the explosively compacted materials, was 89 and 91 K, respectively, whereas the corresponding transition temperature for the initial powder free of silver was 90 K, see Figure 4.36(a). Although there is no significant difference in transition temperatures before and after the explosive powder compaction, a reduction in superconducting volume fraction was indicated by the decrease of the intensity of the Meissner shielding signal, i.e., the magnetization difference or the transition height, ΔM. In addition, the shape of the transition curves became broader after the explosive

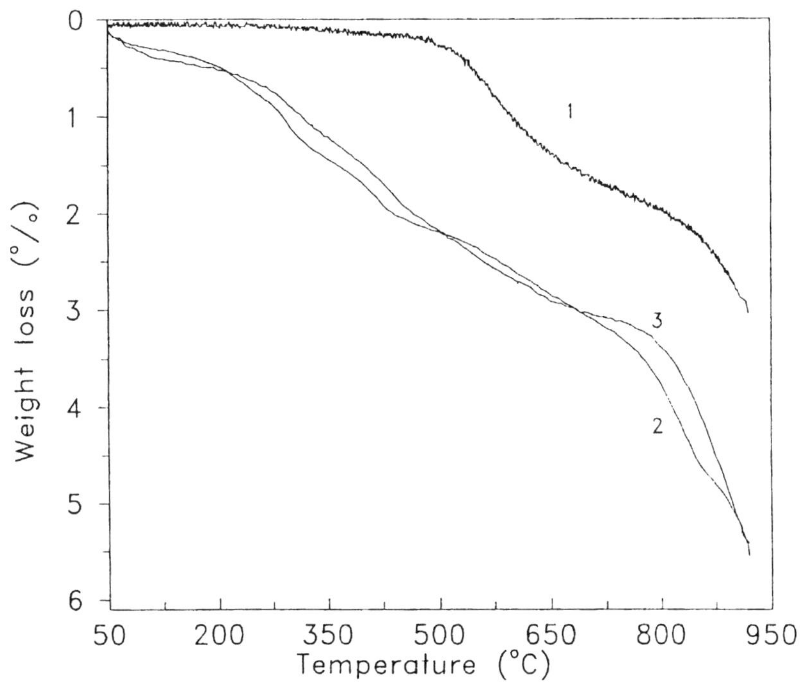

FIGURE 4.35. TGA diagrams. 1: initial Ag-less YBCO powder (before compaction); 2: 1st type compacted powder; 3: 2nd type compacted powder.

processing because of many lattice imperfections and linear defects created by the shock; see Figure 4.36(a).

The critical current density, J_c can be estimated by the hysteresis loop measurements, see Figure 4.36(b). The J_c is given by the Bean critical state model for a specific sample geometry [40]. For rectangular samples of dimensions $w_1 \times l_1 \times h_1$ (w_1 is the width [cm], l_1 the length [cm], h_1 the thickness [cm] of the plate) and for the magnetic field parallel to the $w_1 l_1$ face, the critical current density is calculated by the following relationship:

$$J_c = 20 \frac{\Delta M}{h_1} \tag{4.56}$$

where ΔM is the width of the hysteresis loop (emu.cm^{-3}), corresponding to a certain applied field.

Considering that $\Delta M = 0.7$ emu/g, for an applied field equal to 10 kOe, $\rho = 5.4$ g/cm^3 (85%) and $h = 0.1$ cm and from Equation (4.56) a value of the current density, $J_c = 750$ A/cm^2, is calculated. This value approximates

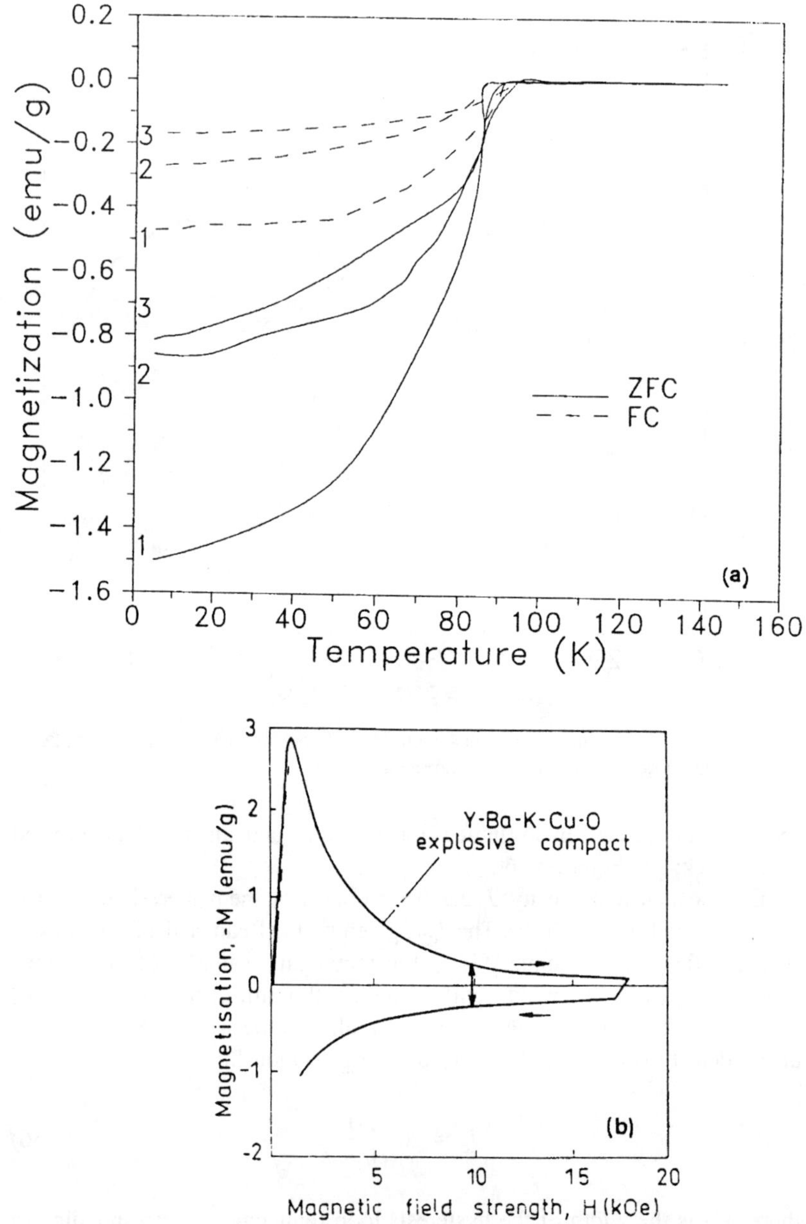

FIGURE 4.36. (a) The dc-magnetization curves. 1: initial Ag-less YBCO powder (before compaction); 2: 1st type compacted powder; 3: 2nd type compacted powder. (b) Hysteresis loop, $M = f(H)$, at 77 K corresponding to the 2nd type compacted powder.

the transport current density obtained from direct J_c measurements. These values of the J_c obtained are rather high for bulk processing and, in particular, for this type of polycrystalline materials that possess a relatively great number of macro- and microdefects, e.g., grain boundaries, cracking, porosity, misorientation, acting as weak links and reducing, therefore, the current carrying capacity of these materials. Note that much higher critical current densities ($>10^4$ A/cm^2) are found in single crystals or thin films.

4.5.3 Explosive Compaction of Axisymmetric Metal/YBCO Billets

Explosive compaction was used for the fabrication of steel/superconductive material/silver composite billets [1,21,22]. The compaction configuration is illustrated in Figure 4.37. Various metals are used as outer powder container: silver (Ag999), plain carbon steel, copper and Cu/Al bimetallic. The setup used for compaction experiments was either the single tube, see Figure 4.37(a), or modified configuration with a madrel situated along the axis of the single-tube setup, see Figure 4.37(b).

The various explosive materials used were:

- The "Nitrammite" high explosive, in granular form, consisting of ammonium nitrate in a percentage above 98%, with 1.0 g/cm^3 bulk density and 4000 m/s detonation velocity; the impact mass ratio $R = 1.4$.
- The "Paxit" high explosive, in granular form, consisting of 77% ammonium nitrate, 19% TNT and 4% w/w inert components, with 1.0 g/cm^3 bulk density, 3800 m/s detonation velocity and impact mass ratio $R = 1.2$.
- The "Nipentex" high explosive, in granular form, with 1.0 g/cm^3 bulk density and a detonation velocity up to 7000 m/s; the impact mass ratio used was equal to 0.5.
- The "Pentaplastite" plastic explosive, consisting of more than 86% PETN and 1.5 g/cm^3 bulk density; the detonation velocity was 7000 m/s, and the impact mass ratio was equal to 1.0.

Explosions were conducted in sand by firing a fuse. A short detonating cord, connected with the fuse, was placed around the cylinder to create a uniform ring-like detonation wave. After detonation, the stress wave generated from the ultra rapid gas release, resulted in the collapse of the steel container and the consolidation of the powder. The bulk density of the ceramic before compaction was up to 40–45%, whereas the bulk density after the compaction was increased to about 85% of the nominal density. The detonation of "Pentaplastite" leads to the formation of very strong shock waves, resulting in the catastrophic rupture of the metal container and in powder evacuation, see Figure 4.38. From the experimental observations made, it may be calculated that the "Paxit" and

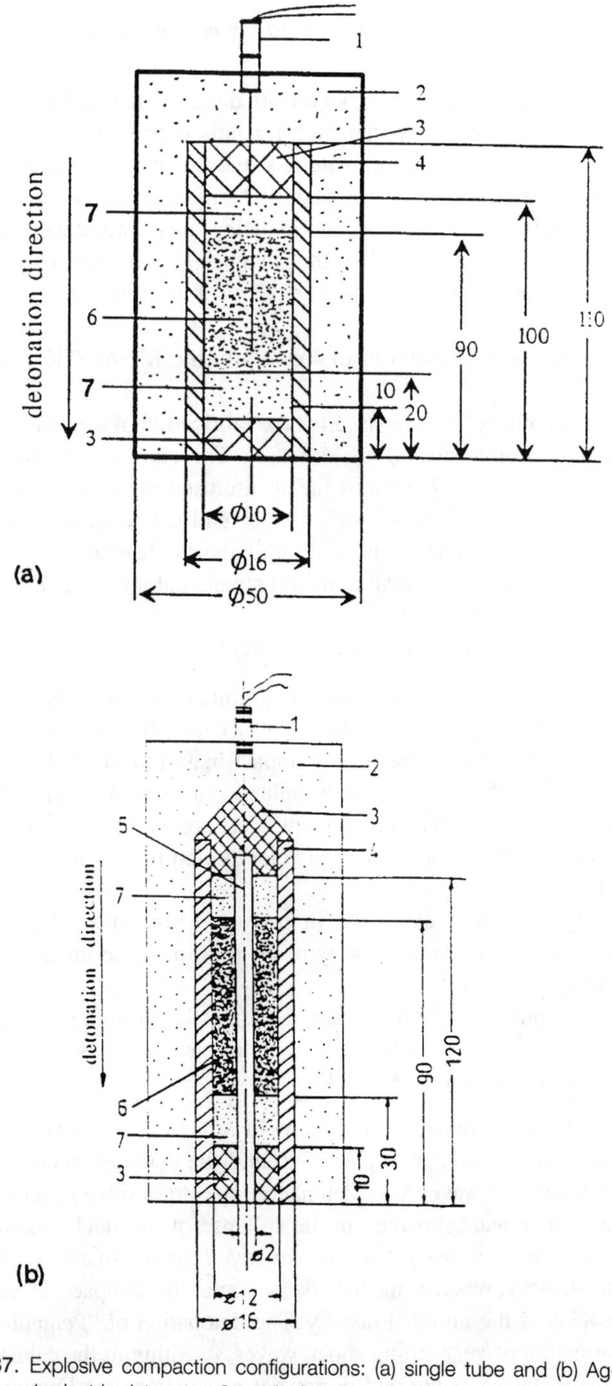

FIGURE 4.37. Explosive compaction configurations: (a) single tube and (b) Ag-mandrel modified setup. 1: electric detonator; 2: explosive; 3: end plug; 4: metal container; 5: mandrel; 6: YBCO powder; 7: MgO powder (dimensions in mm).

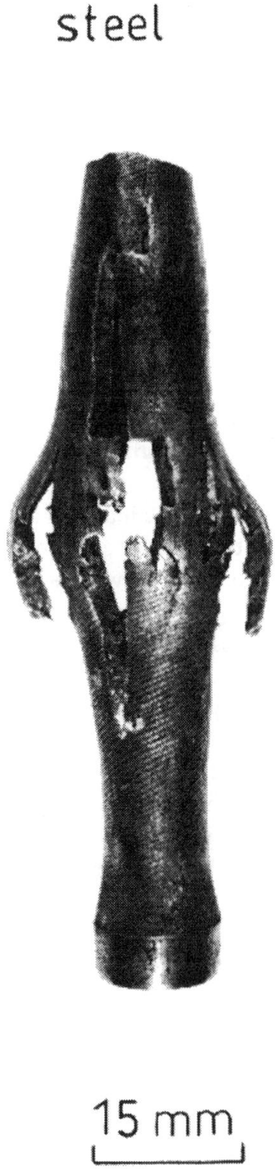

FIGURE 4.38. Catastrophic burst of the steel container of the explosive compaction chamber after Pentaplastite detonation.

"Nitrammite" explosives were optimal for the explosive compaction conditions used. On the other hand, the "Nipentex" explosive results in successful powder consolidation only when used in small impact mass ratios.

From Equation (4.55) the following compaction parameters are calculated: $v_p = 970$ m/s; $P = 6.5$ GPa and $E = 4.5 \times 10^5$ J/kg, when an Ag container was used; $P = 9.4$ GPa and $E = 6.5 \times 10^5$ J/kg, when a steel container was used.

After compaction the billets were surface finished by turning to remove the burnt outside surface, see Figure 4.39. Longitudinal and transverse sections were then cut off and prepared properly for microscopic examination.

Optical microscopy reveals a very dense microstructure, see Figure 4.40, and a good bonding of the metal sheath/ceramic interface, see Figure 4.41(a). A micrograph showing parallel microcracks perpendicular to detonation direction, which are formed because the tensile stress acting on the ceramic core, is presented in Figure 4.41(b). The microstructure of the ceramic obtained by SEM is shown in Figure 4.42.

The examination of the steel/ceramic interface and its vicinity, by scanning electron microscopy, revealed two discrete zones of their morphology, see Figure 4.43(a). *Zone 1,* adjacent to the metal wall, is around 40-μm thick and relatively homogeneous, consisting of fine grains that resulted from the intense grain fragmentation and the possible partial melting at the region near the interface. *Zone 2* starts at a distance, about 40 μm, from the steel wall; its morphology is inhomogeneous, compared with that of Zone 1, consisting of grains of various dimensions, with a larger average grain size. Note that this morphology is similar to the microstructure consisting of coarse grains, dispersed in a fine-grained matrix, which is the dominant compacted ceramic microstructure; see 1 and 2 in the same Figure. EDS analysis revealed that no change in chemical composition between these two zones occurred; the chemical composition of the individual grains of different shape and dimensions was almost the same, see Figure 4.44(a).

Examination of the silver/ceramic interface revealed relatively good bonding characteristics, but without any zone differentiation.

A macroscopic cross section, composed of a circular sector of the silver mandrel, is shown in Figure 4.43(b); details of the silver/ceramic interface are clearly indicated in photograph 1 of Figure 4.43(b). Localized vortices may be formed at the silver/ceramic interface that resulted from process hydrodynamics, see photograph 2 in Figure 4.43(b).

X-ray diffraction patterns of the compacted ceramic revealed the presence of the orthorhombic high-T_c superconducting phase, whereas the relatively weaker and broader diffraction peaks, compared with those of the initial powder, indicate significant reduction of the grain size and lattice distortion that resulted from dynamic compaction; see Figure 4.44(b). Various types of im-

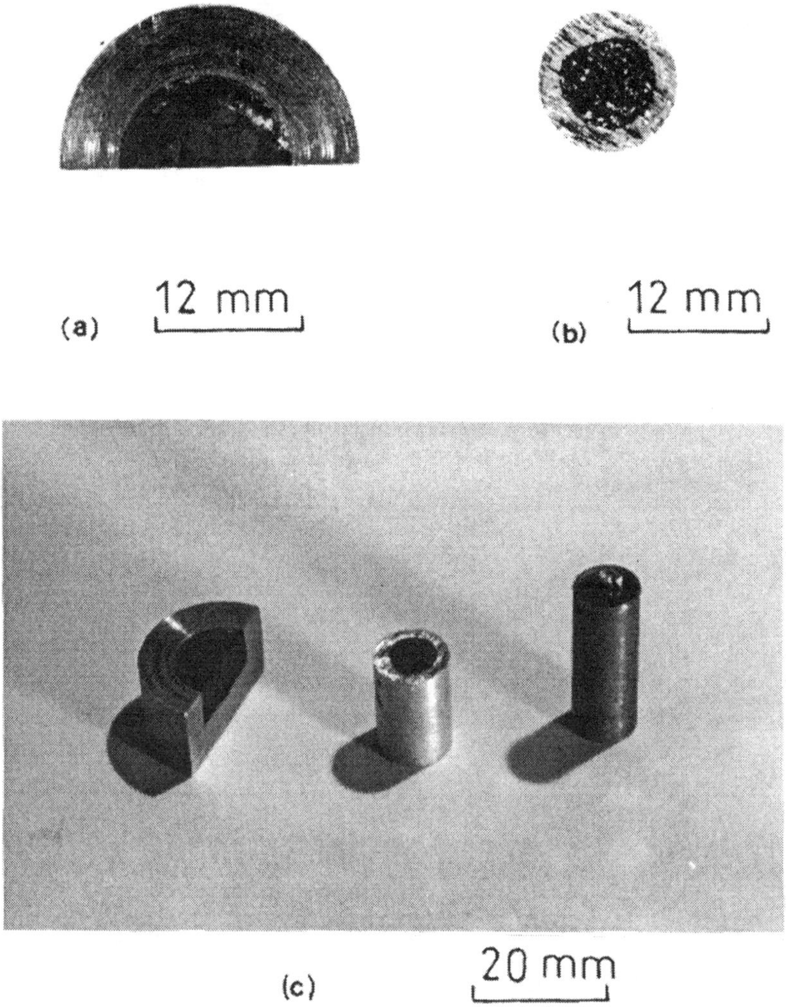

FIGURE 4.39. Metal sheathed YBCO billets fabricated by explosive compaction.

purities, in low percentage, such as $BaCuO_2$ and Y_2BaCuO_5, formed probably because of local decomposition of the high-T_c compound, were also detected.

Several microstructural defects were observed in the bulk compacted ceramic by using optical and scanning electron microscopy. Aggregates of ce-

FIGURE 4.40. Optical micrographs showing (a) the microstructure of the compacted YBCO powder and (b) detail of (a) (Nitrammite explosive, R = 14).

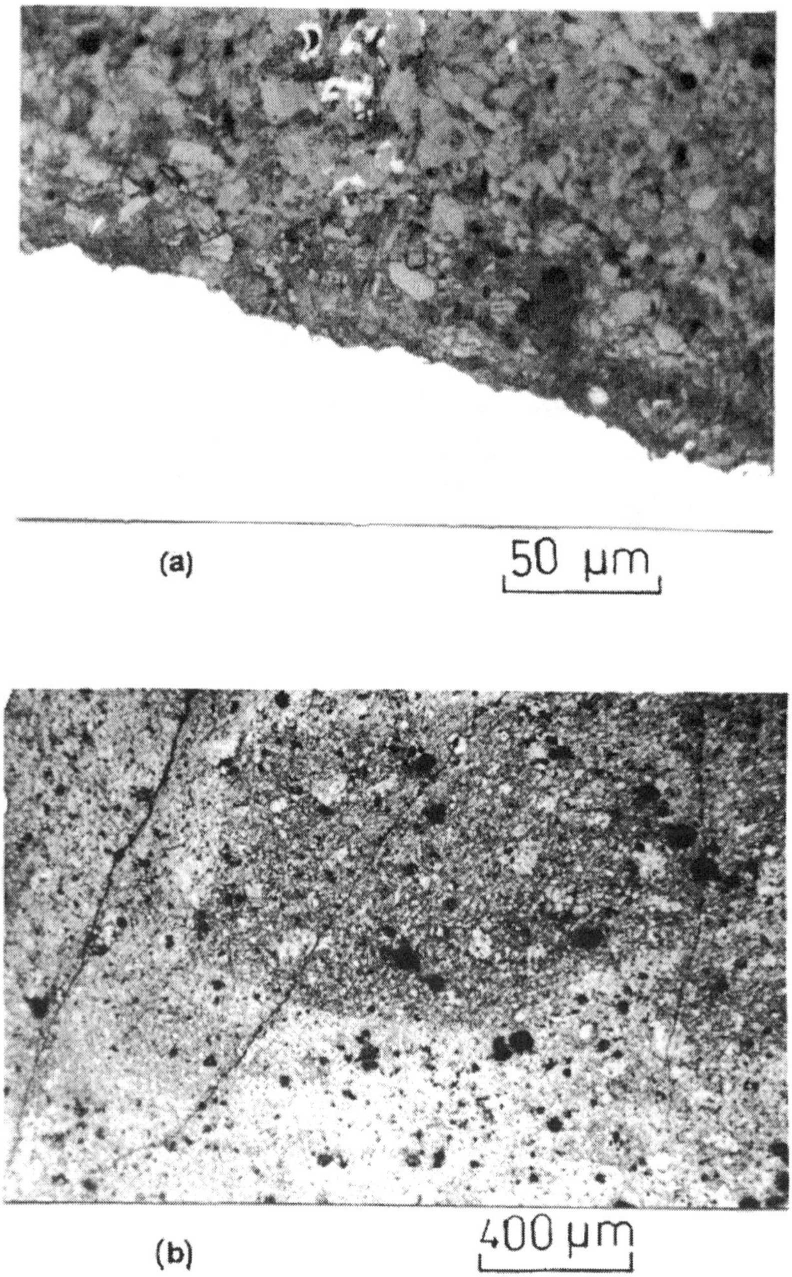

FIGURE 4.41. Optical micrographs showing (a) the microstructure near the steel/ceramic interface and (b) the presence of parallel microcracks (perpendicular to detonation direction) of explosively compacted YBCO ceramic. (Nitrammite explosive, R = 1.4).

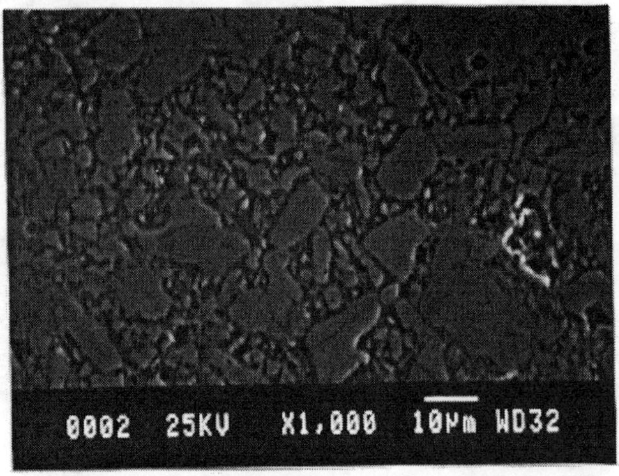

(a)

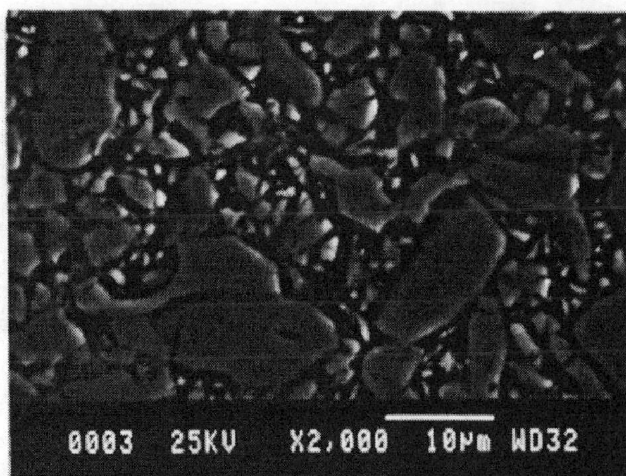

(b)

FIGURE 4.42. Scanning electron micrographs showing (a) the microstructure of the explosively compacted YBCO ceramic and (b) detail of (a). (Nitrammite explosive, R = 1.4).

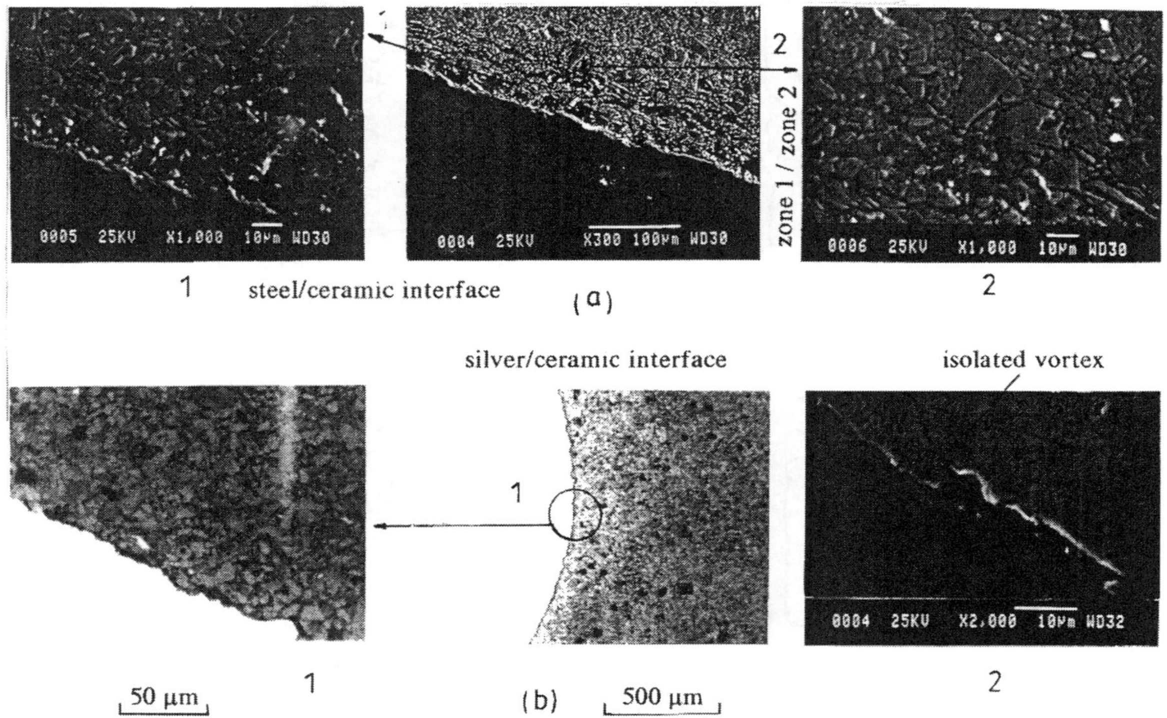

FIGURE 4.43. Micrographs showing (a) the microstructural changes near the metal container steel/YBCO ceramic interface; 1: detail of zone 1; 2: detail of zone 2 and (b) a cross section at the mandrel (Ag)/ceramic interface; 1: detail of the microstructure near the interface; 2: localized vortex formation.

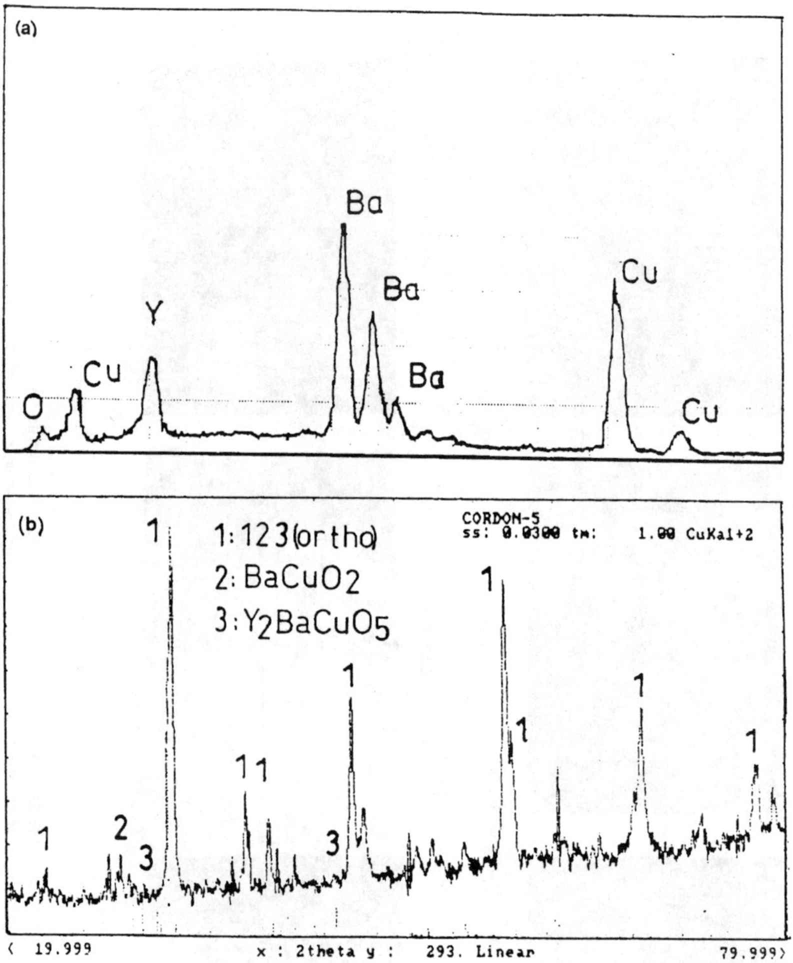

FIGURE 4.44. (a) EDS spectrum and (b) XRD pattern of the explosively compacted YBCO ceramic.

ramic material that resulted from grain coalescence were observed at sites of possible superposition of shock waves, leading to high-pressure development and local temperature rise; see Figure 4.45(a). Individual grain fracturing is a common phenomenon in explosive compaction, leading to the formation of intragrain boundaries, which result a substantial reduction of the porosity and in the enhancement of the critical current density; see Figure 4.45(b). Needle-like grains, corresponding to barium-based compounds, and in particular to barium cuprate ($BaCuO_2$), were detected by high-magnification scanning elec-

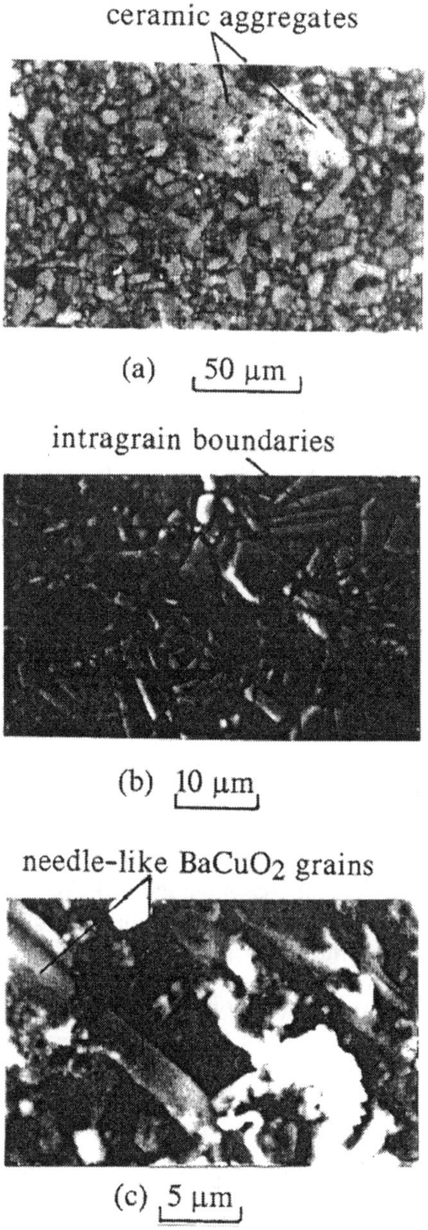

FIGURE 4.45. Microstructural features of explosively compacted YBCO ceramic: (a) ceramic aggregates; (b) intragrain boundaries; (c) needle-like crystals corresponding probably to $BaCuO_2$ phase. (Nitrammite explosive, R = 1.4).

tron microscopy, as shown in Figure 4.45(c); these types of impurities were also previously reported, see Reference [2].

The metallographic examination of the polished sections revealed the presence of various types of shock-induced microcracks, see Figure 4.46 and Section 4.4.6:

- Microcracks perpendicular to the axis of symmetry: These microcracks that resulted from the longitudinal tensile stresses propagate perpendicular to the applied stress direction, see 1 of Fig. 4.46. It may be designated as a Mode I type of fracture, i.e., tensile crack-tip opening.
- Microcracks developed at $\pm 45°$ to the axis of symmetry: These are mainly shear cracks, and their direction coincides with the maximum shear stress direction (CRSS); see 2 of Figure 4.46.
- Interfacial microcracks: These microcracks, developed in dynamic compaction or in explosive welding, are the spall fractures due to the impedance mismatch and to the subsequent reflection of strong tensile stress waves, see 3 of Figure 4.46.
- Crack deflection: An example of this type of failure is shown in 4 of Figure 4.46. It is probably caused by a fracture Mode transition; from Mode I (crack-tip opening) to Mode II (shearing).
- Crack-branching (bifurcation): This is a characteristic fracture observed, when ultra-high explosives, such as Nipentex, are used; two different crack branches are generated from a primary crack. Crack-branching is an example of unstable dynamic crack propagation, analyzed by Yoffé [41]. According to Yoffé's treatment, crack-branching occurs, when the strain energy release rate parameter, G, is higher than the material fracture resistance, R_f. The periodicity of multiple crack-branching phenomena is shown in the diagram presenting the variation of G as a function of the increase of crack length, $\Delta\alpha$, see Figure 4.47(a); a characteristic microcrack pattern of bifurcation is shown in Figure 4.47(b).

The evolution of the real part, χ' of the ac-susceptibility as a function of temperature, at the various stages of fabrication, is presented in Figure 4.48. The normal/superconducting transition is relatively sharp, with an onset critical temperature up to 93 K, whereas the corresponding transition width is around 13 K, as shown in Table 4.2. The shift of the transition to slightly higher onset critical temperature, compared with the initial powder, may be attributed to:

- the possible contraction of the perovskite unit cell due to dynamic compaction
- the redistribution of the crystal defects in the high-pressure state, leading, therefore, to the normalization of the existing oxygen vacancies

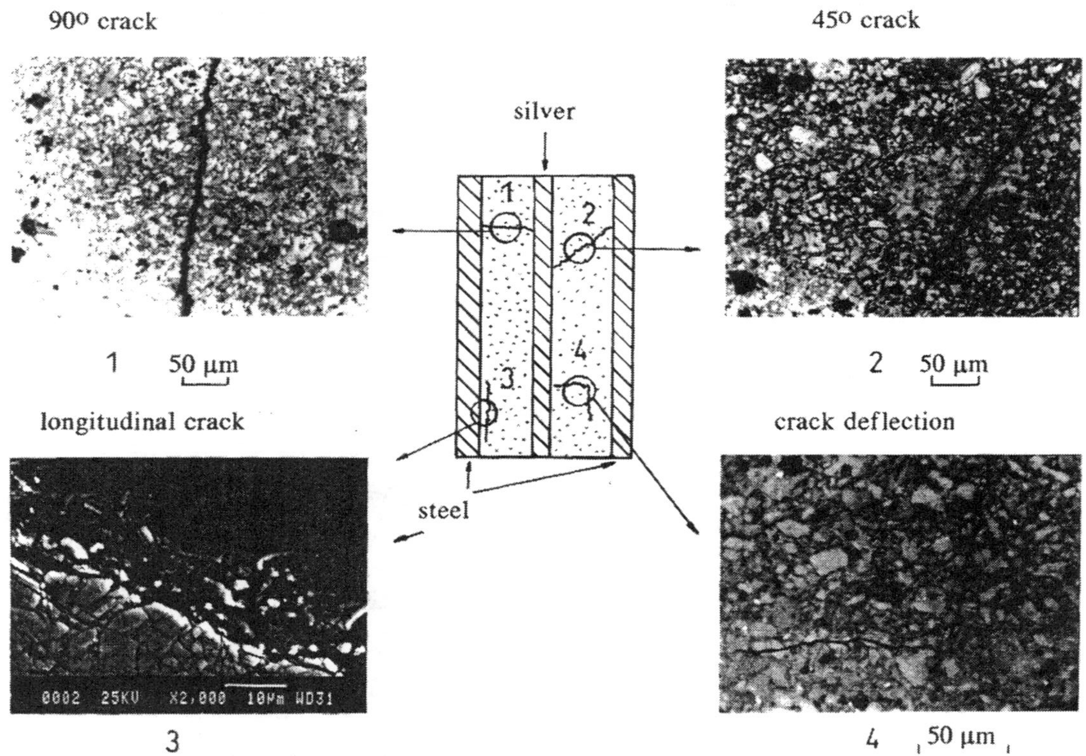

FIGURE 4.46. Micrographs showing various types of shock-induced microcracks during the explosive compaction of YBCO ceramics. 1: 90° crack; 2: 45° crack; 3: longitudinal (interfacial) crack; 4: crack-deflection. (Nitrammite explosive, R = 1.4).

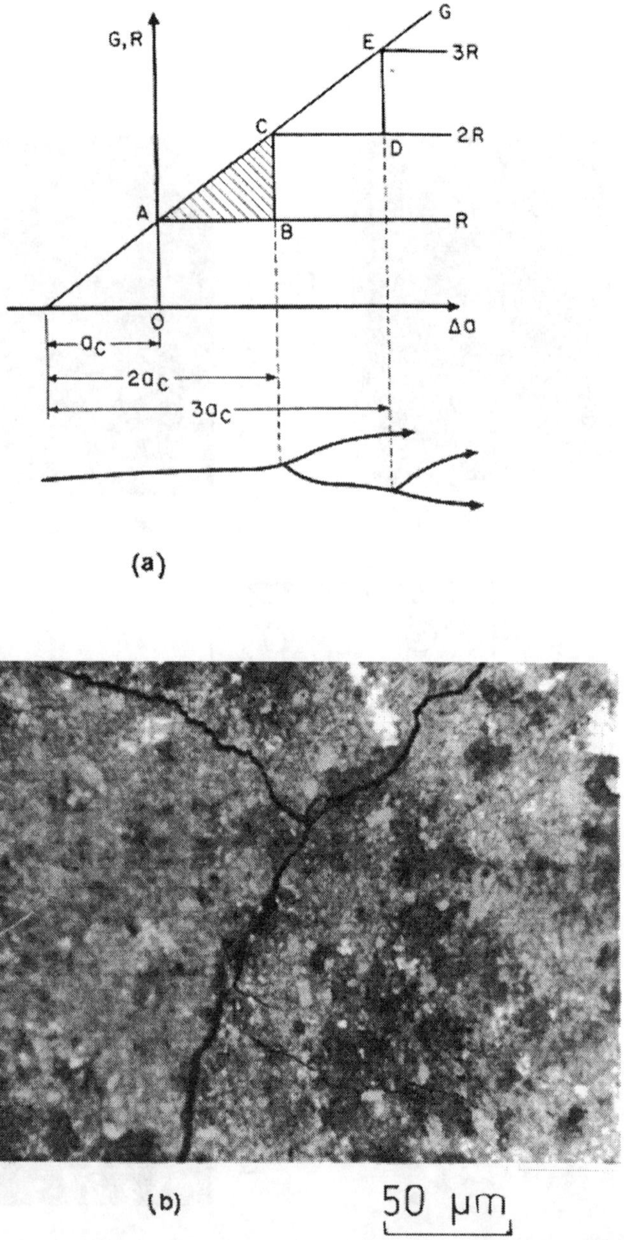

FIGURE 4.47. (a) Variation of strain release parameter, *G*, and fracture resistance, *R*, with $\Delta\alpha$ (from Reference [41]). (b) Optical micrograph showing a bifurcation formed because of unstable crack propagation for explosively compacted YBCO ceramic. (Nipentex explosive, *R = 0.5)*.

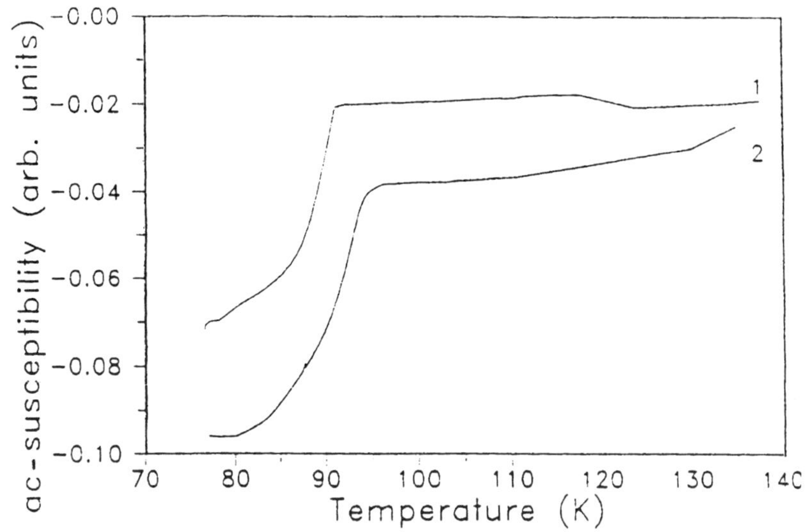

FIGURE 4.48. Variation of ac-magnetic susceptibility (χ') as a function of temperature. 1: initial YBCO powder; 2: explosively compacted powder.

4.5.4 Thermal Treatment of Explosively Compacted YBCO Billets

Ceramic powder mixture corresponding to the stoichiometric ratio of $YBa_{1.95}K_{0.05}Cu_3O_y$ superconducting ceramic with 10% w/w Ag was heat-treated at 950°C for 3 h in flowing oxygen. The resultant preheat-treated ceramic superconducting powder was consolidated in a silver tube by explosive compaction technique. After compaction, the preheat-treated billets were subjected to annealing in oxygen stream with an oxygen flow rate up to 15 dm^3/h, by heating up at 850°C for 10 h and then slowly cooling at room temperature; the heating/cooling rates were equal to 300°C/h, see Figure 4.49(a) and Reference [42].

The same fabrication technique, i.e., explosive compaction, was used for the initial stoichiometric raw materials mixture, without prior heat treatment, and then the "green" product was subjected to a series of heat-treating cycles (920°C, 70 h) in oxygen stream, to obtain the required superconducting properties (postheat-treated ceramic). Note that as before, the oxygen flow rate was

TABLE 4.2. Superconducting Properties of YBCO before and after Explosive Compaction.

Material	$T_{c,onset}$ (K)	$T_{c,offset}$ (K)	ΔT_c (K)
Initial powder	90	81	9
Compacted ceramic	93	80	13

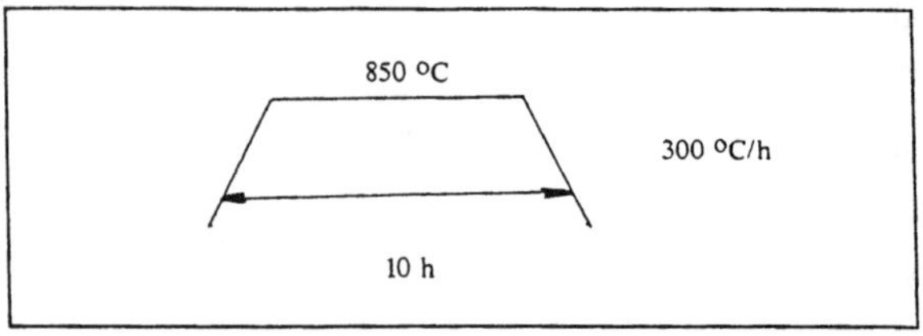

(a)

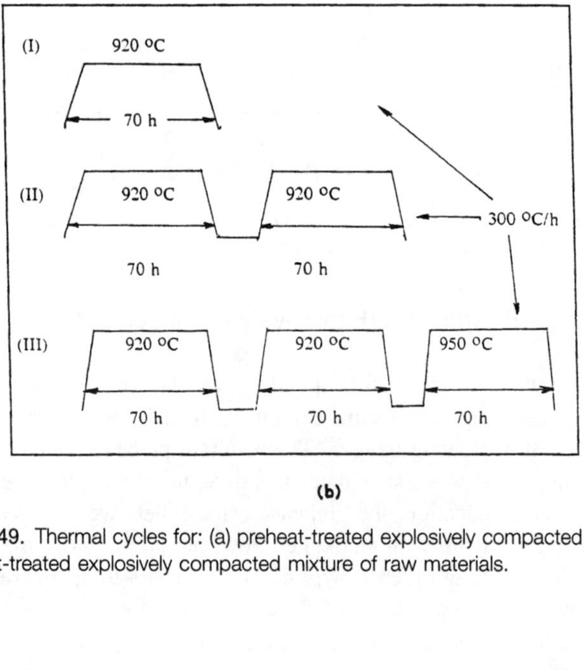

(b)

FIGURE 4.49. Thermal cycles for: (a) preheat-treated explosively compacted powder and (b) postheat-treated explosively compacted mixture of raw materials.

up to 15 dm^3/h, and the cooling/heating rates were equal to 300°C/h. The various thermal cycles used for the heat treatment are shown in Figure 4.49(b); see also Reference [42].

The microstructure and the physical properties of the preheat- and postheat-treated ceramics were characterised and compared, to examine the effect of heat treatment on the soundness and the quality of these advanced ceramic compacts.

PRE-HEAT TREATED EXPLOSIVELY COMPACTED CERAMIC BILLETS

The heat-treated pellets, before compaction, show a characteristic sintered microstructure consisting of plate-like YBCO grains, see Figure 4.50. Firing

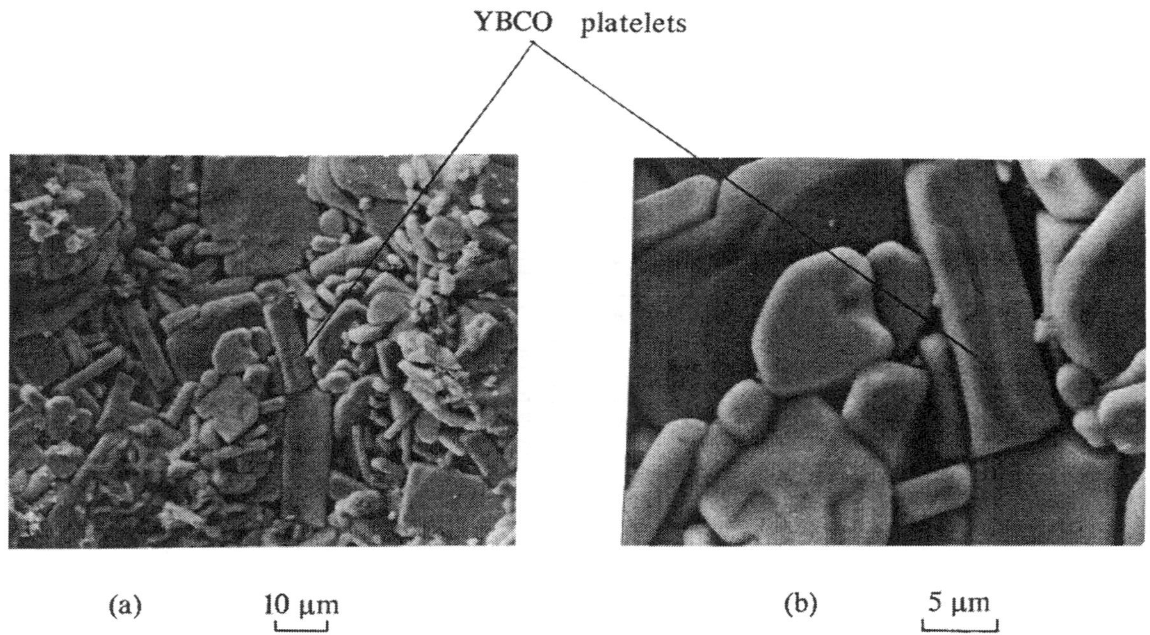

FIGURE 4.50. Scanning electron micrographs showing (a) the microstructure of the sintered YBCO ceramic and (b) a detail of (a).

at 950°C, besides chemical reaction, activates solid-state diffusion, leading to "neck" formation among crystals. The pore volume parameter was measured by using mercury porosimetry technique and found equal to 7.5×10^{-2} cm^3/g corresponding to a bulk density of 65% of the theoretical one.

The addition of KF in the raw materials mixture aims to the partial replacement of Ba in the crystal lattice of the 123 compound. Doping the Y-Ba-Cu-O ceramic with alkali cations (Li, Na, K) favors the stabilization of 123 orthorhombic phase ($YBa_2Cu_3O_{7-\delta}$) at lower sintering temperatures and durations; the sintering duration performed in this work was 3 h. This can be explained by the comparative XPS study of nondoped and K-doped samples; in the latter case a greater amount of trivalent copper ions was found according to the following reaction [43]:

$$Cu^{2+} + Ba^{2+} \rightarrow K^{+} + Cu^{3+}$$

Potassium atoms form easier a superconducting solid solution by partial barium substitution $Y(Ba,K)_2Cu_3O_{7-\delta}$ than other alkali atoms (K > Na > Li). This is due to the position of the above alkali elements in the periodic table, relative to barium, explaining, therefore, the potassium greatest compatibility with barium, which has the nearest atomic radius (0.227 nm) compared to barium (0.217 nm) and the same coordination number (CN = 6). On the other hand, oxygen ions can be partially replaced by fluoride, which is a more electronegative atom, reducing the sensitivity of the YBCO compounds to oxygen content. These two substitutions, i.e., Ba and K and O by F, result in an increase of T_c by 6–10 K and of J_c at 77 K; see References [44,45].

The XRD pattern of the doped compound, after the above-mentioned heat treatment, shows a content of secondary phases (Y_2BaCuO_5, CuO and $BaCuO_2$) up to 5% w/w, Ag 10% w/w and 123 phase 85%, see Figure 4.51(a). The T_c of the sintered materials, measured from the dc-magnetic susceptibility curve, $\kappa = f(T)$, was found equal to 102 K, see 1 in Figure 4.52.

The explosively compacted ceramic shows a profound increase of density due to intense particle fracturing, void collapse and microwelding and bonding at the grain boundaries that resulted from the shock wave propagation through the porous medium. The pore volume parameter measured by a mercury porosimeter was found equal to 2.9×10^{-2} cm^3/g, which corresponds to 85% bulk density. The shock wave passage results in a considerable microstructural refinement, leading to the formation of a near-nanocrystalline morphology, e.g., grain size less than 1 μm and to the creation of lattice distortion and/or incompleteness in crystallinity, see Figure 4.53(a) and (b). The EDS spectrum of the compacted preheat-treated ceramic shows a stoichiometric mol ratio Y:Ba:Cu = 1:2.39:3.20, see Figure 4.54. Qualitative detection of the microcrystallinity and the shock-induced lattice defects can be pro-

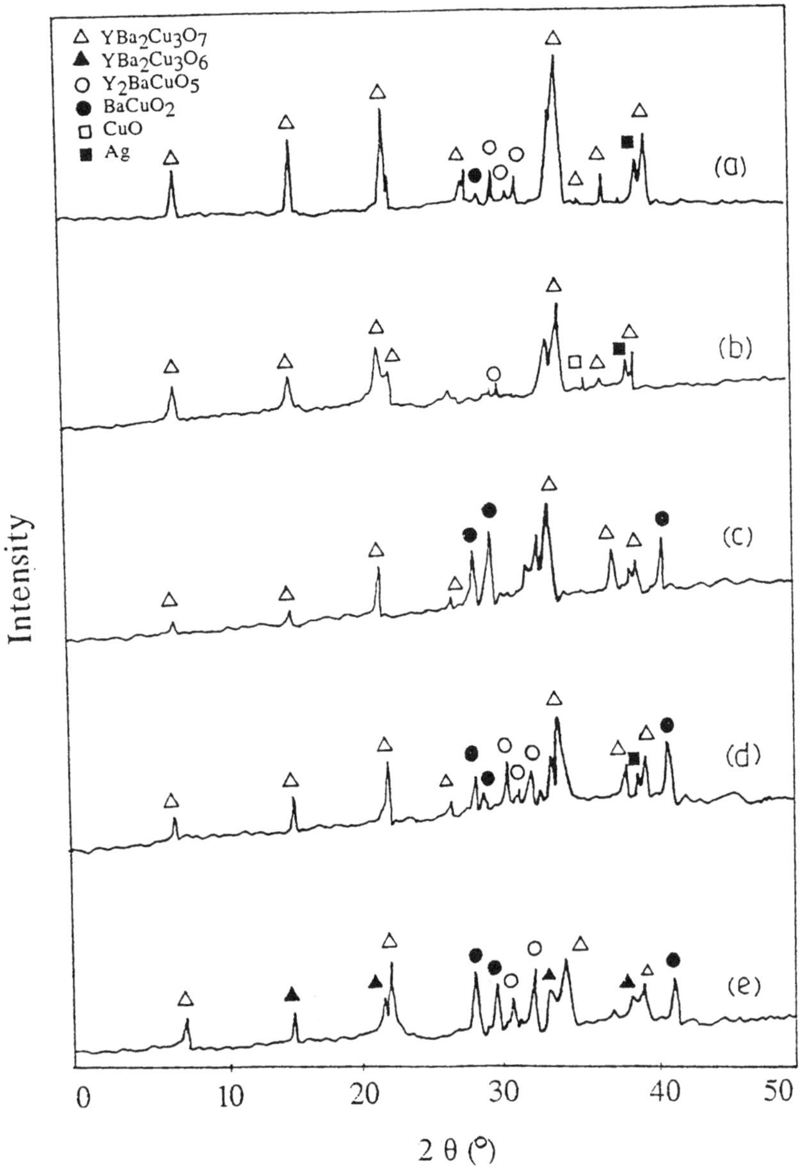

FIGURE 4.51. XRD patterns of the (a) sintered YBCO; (b) preheat-treated explosively compacted ceramic; (c) postheat-treated explosively compacted ceramic (920°C, 70 h); (d) postheat-treated explosively compacted ceramic (920°C, 70 h/920°C, 70h); (e) postheat-treated explosively compacted ceramic (920°C, 70 h/920°C, 70h/950°C, 70 h).

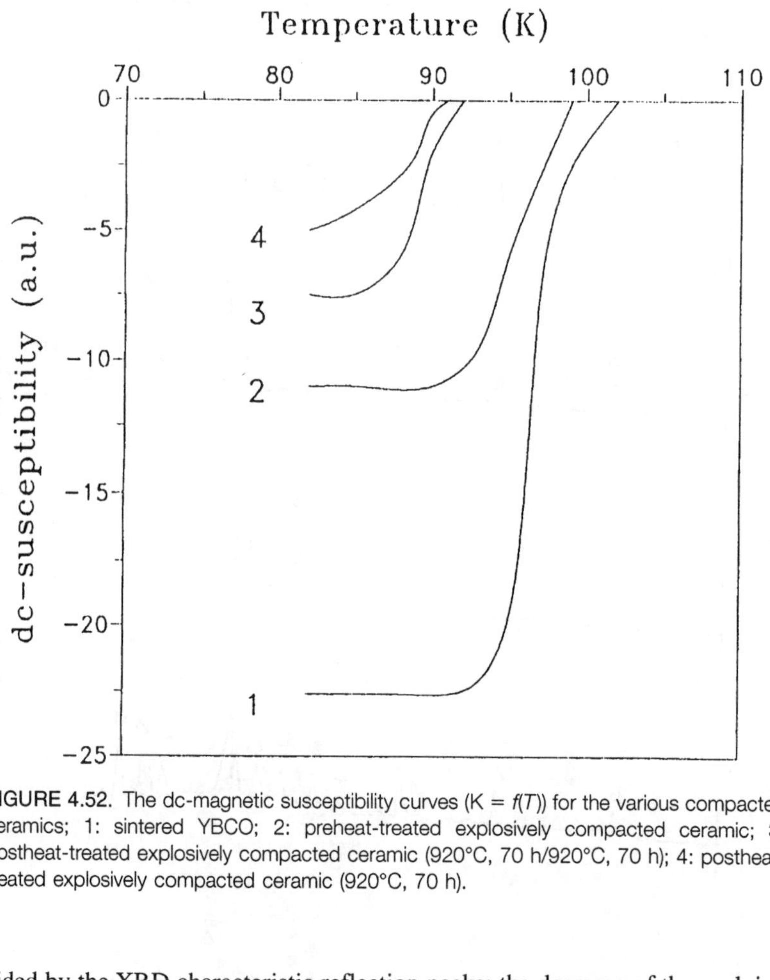

FIGURE 4.52. The dc-magnetic susceptibility curves (K = $f(T)$) for the various compacted ceramics; 1: sintered YBCO; 2: preheat-treated explosively compacted ceramic; 3: postheat-treated explosively compacted ceramic (920°C, 70 h/920°C, 70 h); 4: postheat-treated explosively compacted ceramic (920°C, 70 h).

vided by the XRD characteristic reflection peaks; the decrease of the peak intensity and the increase of peak width, shown in Figure 4.51(b), indicates this fact. Note also that a small decrease of the "green" phase Y_2BaCuO_5, accompanied by the disappearance of $BaCuO_2$, was observed from the XRD pattern, see also Table 4.3. This may be explained by the incogruent melting, during compaction, of a very small percentage of the 123 phase, including the dissolution of $BaCuO_2$, at temperatures above 970°C; the subsequent rapid cooling may lead to the increase of the Y_2BaCuO_5 phase.

The dc-magnetic susceptibility curve of the explosively compacte preheated ceramic shows the characteristic Meissner effect at T_c = 99 K, see 2 in Figure 4.52. The relative decrease of the diamagnetic shielding signal, compared with the sintered sample, is related to the decrease of the percentage of the 123 or-

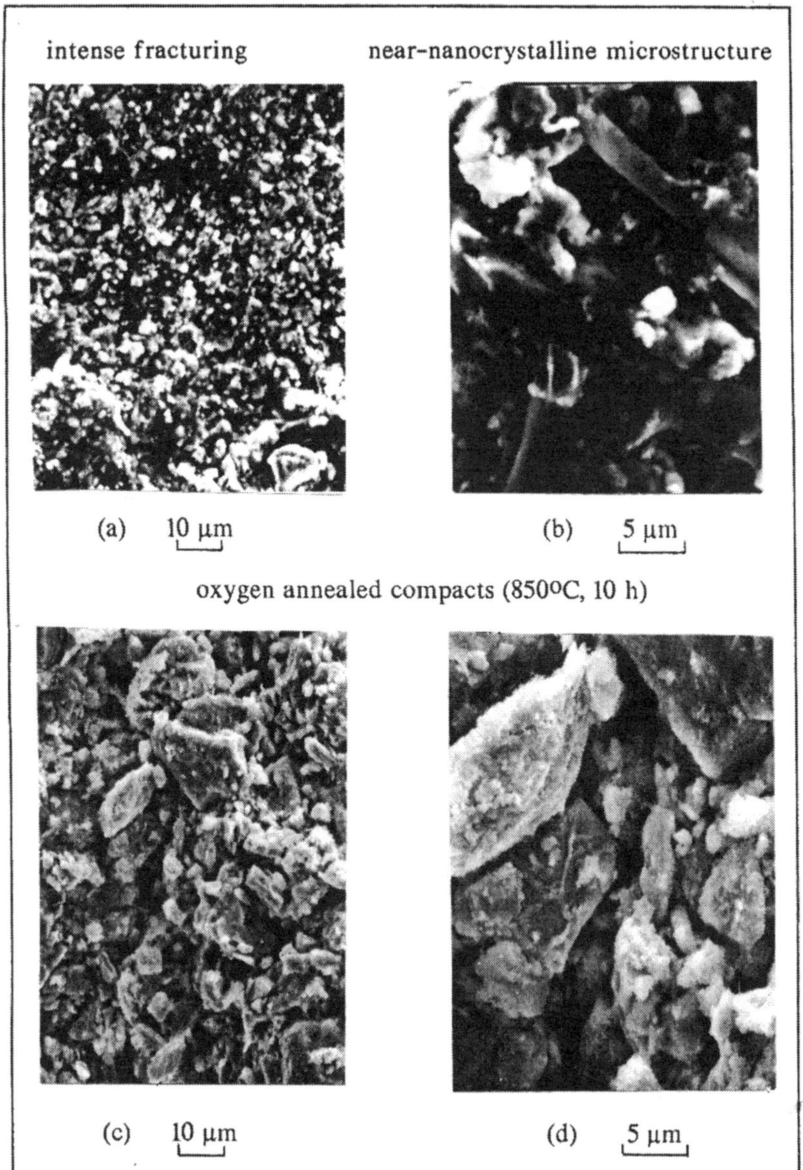

FIGURE 4.53. Scanning electron micrographs showing (a) the cross-section of the preheat-treated explosively compacted YBCO ceramic billet; (b) a detail of (a); (c) the microstructure of the preheat-treated compact after subsequent oxygen annealing at 850°C for 10 h; (d) a detail of (c). The detonation direction is vertical to the paper plane.

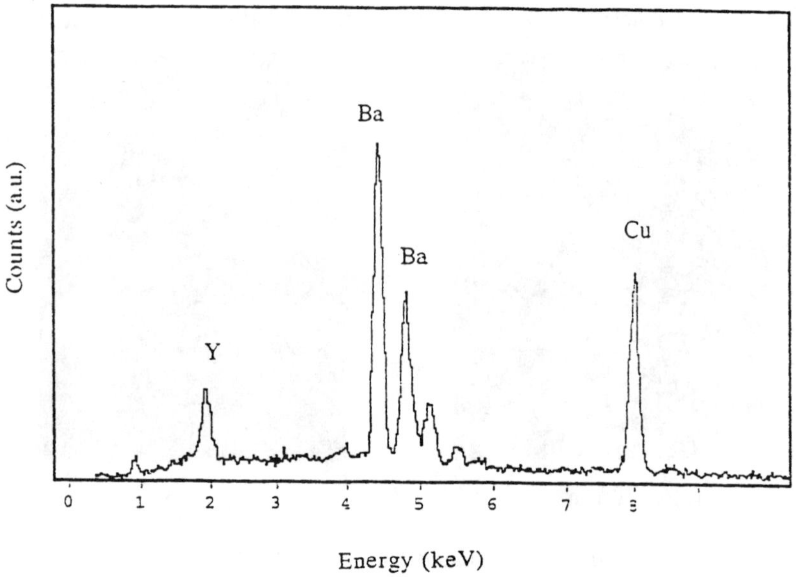

FIGURE 4.54. EDS spectrum corresponding to the preheat-treated "green" YBCO compact.

thorombic phase. The obtained critical current density, J_c for the compacted preheated ceramic is 800 A/cm^2 at 77 K; the presence of secondary phases in the grain boundaries (weak links) and of microcracks observed in the compacted ceramic, in relation also to the porosity of the compact, may be associated with the relatively low values of J_c obtained. Higher critical current densities can be obtained after annealing of the compacted ceramic in oxygen stream at 850°C for 10 h and a flow rate of 15 dm^3/h. The crystallization of the 123 phase can be controlled by heating in oxygen flow, and it is influenced by the solid-state diffusion processes that result in grain growth and, therefore, in pore and microcrack elimination. The measured pore volume was less than 0.01 cm^3/g, which corresponds to a bulk density greater than 94%. The considerable grain growth resulted in microstructures of grains 5–10 μm, see Figure 4.53(c) and (d). In addition, higher grain sizes may lead to greater supercurrent loops, increasing, therefore, the pinning forces and, consequently, the intragrain current density. The J_c value measured for the annealed HTS sample was 2800 A/cm^2 at 77 K.

POST-HEAT TREATED EXPLOSIVELY COMPACTED CERAMIC BILLETS

Further treatment is necessary for the nonheat-treated explosively compacted Ag/YBCO billet to generate superconductivity in the consolidated ce-

TABLE 4.3. Phase Content of the Explosively Compacted YBCO Ceramics (Number of Counts per Second).

Compact Phase	Preheat-Treated (950°C, 3 h)	Preheat-Treated (920°C, 70 h)	Preheat-Treated (920°C, 70 h/920°C, 70 h)	Preheat-Treated (920°C, 70 h/920°C, 70 h/950°C, 70 h)
$YBa_2Cu_3O_7$	5084	1116	2992	1253
$YBa_2Cu_3O_6$	—	—	—	1122
Y_2BaCuO_5	49	—	404	853
$BaCuO_2$	—	992	713	1119
CuO	66	—	—	—

ramic. The selection of the heating temperature is based on the primary conventional reaction processing, applied to YBCO superconducting powders, and, particularly, it must be between 900 and 960°C (silver melting temperature); see above. The heat-treating duration must be long because of the limited surface energy provided by the compact shape of the ceramic core. The presence of silver increases the oxygen diffusivity from the external metal container to the core and accelerates the diffusion processes in the compacted body, "wetting" the ceramic grains and creating effective diffusion and reaction channels.

The XRD patterns of the postheat-treated ceramic, with respect to the single- and double-thermal cycles applied after the explosive compaction [see (I) and (II) in Figure 4.49(b)] are shown in Figure 4.51(c) and (d); a tendency of the 123 phase formation is indicated; enhancement of 123 phase content is observed in the case of double-thermal cycle heat-treated compacted billet, see also Table 4.3. Further heat treatment, including an additional third step at 950°C for 70 h [see III in Figure 4.49], may lead to partial transformation of the 123 orthorhombic phase $YBa_2Cu_3O_7$ to the tetragonal one $YBa_2Cu_3O_6$ and to secondary phases Y_2BaCuO_5 and $BaCuO_2$, see Figure 4.51(e) and Table 4.3; a subsequent decrease of the residual superconductivity of the compact is apparent. The microstructure of the optimum, as far as the superconductivity is concerned, postheat-treated ceramic compact subjected to double-thermal cycle (920°C for 70 h/920°C for 70 h) is shown in Figure 4.55. The solid-state diffusion, which took place during prolonged heat treatment, has led to excessive grain growth, i.e., grain sizes 15–30 μm, similar to secondary recrystallization observed in metals, see Figure 4.55(b), resulting in strong interparticle bonding and porosity elimination, see Figure 4.55(c); bulk densities greater than 97% were obtained.

The critical transition temperatures for the single- and double-thermal cycle postheat-treated compact are 91 and 92 K, respectively, as obtained from the dc-magnetic susceptibily curves, see 4 and 3 in Figure 4.52; the diamagnetic signal is slightly increased in the second case, showing the relative increase of the 123 superconducting phase content mentioned above. Note, however, that for the diamagnetic properties, the preheat-treated compacted ceramic exhibits a better behavior than the postheat-treated one; compare 2 and 3 in Figure 4.52.

4.5.5 Compaction/Cladding of Metal/BSCCO Plates

Explosive compaction/cladding was used for the fabrication of the BSCCO superconductor/silver composite plates for producing strips by subsequent cold rolling [46]. A cladding configuration used is shown schematically in Figure 4.56. The explosive used was the "Paxit," consisting of about 77% ammonium nitrate,

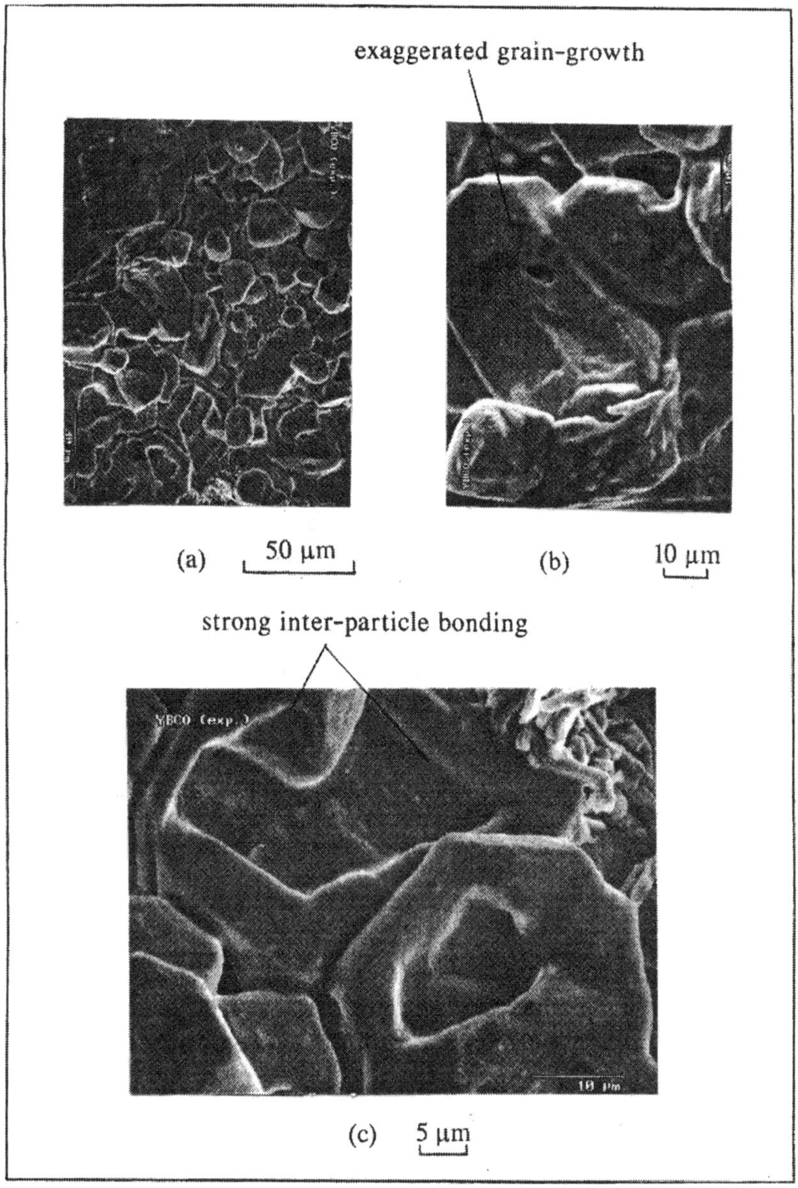

FIGURE 4.55. Scanning electron micrographs of the preheat-treated explosively compacted ceramic (920°C, 70 h/920°C, 70 h), showing (a) the microstructure of a fractured cross section; (b) a detail of (a); (c) the strong interparticle bonding occurred during heat treatment.

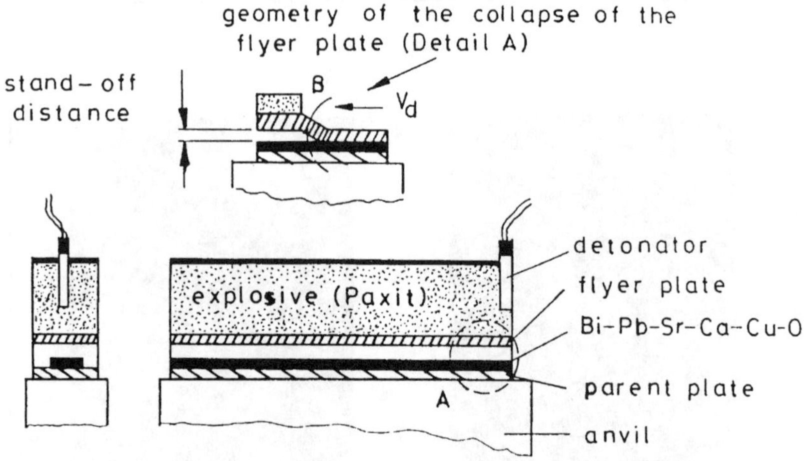

FIGURE 4.56. Experimental 3D compaction/cladding setup for BSCCO ceramic.

19% TNT and 4% inert components, in granular form, with 1.0 g/cm^3 bulk density and 3800 m/s detonation velocity. An explosive mass of 200 g was used, forming a layer of 25-mm thickness above the flyer plate, as shown in Figure 4.56. The impact mass ratio R was 1.2, the initial setup angle 0° and the stand-off distance 2 mm. The bulk density of the powder, before explosive cladding, was about 50% of the nominal density (theoretical density) and the final bulk density, after explosive cladding, was about 85% of the nominal density. Explosions were conducted in sand by firing an electric detonator, resulting in the collapse of the flyer plate and the consolidation of the powder. The explosive cladding parameters used, calculated from Equation (4.54), are tabulated in Table 4.4.

TABLE 4.4. Explosive Compaction/Cladding Parameters (BSCCO).

Dimension of flyer plate (Ag)	200 × 40 ×2 (mm)
Dimension of parent plate (Ag)	200 × 40 × 2 (mm)
Type of explosive	Paxit
Detonation velocity	3800 (m/s)
Initial setup angle	0 (°)
Collision angle	13 (°)
Standoff distance	2 (mm)
Thickness of explosive	25 (mm)
Impact mass ratio	1.2
Velocity of flyer plate	1047 (m/s)
Peak shock pressure in Ag/Ag	16 (GPa)
Peak shock pressure in HTS (2)	10 (GPa)
Kinetic energy of flyer plate	7.4×10^7 (J/m^2)
Increase of internal energy HTS(2)	6.2×10^5 (J/kg)

Microscopic observations of the explosively cladded plates, by using optical microscopy, reveal several macro- and microdefects. Poor adhesion and partial detachment of the superconducting layer from the base (parent) plate were observed, mainly because of the reflection of the shock waves at the interface between the high-T_c ceramic and the silver plate, where a great impedance change is expected , see Figure 4.57(a) and (b).

Microcracks, almost perpendicular to the cladding direction, were observed in the superconducting material, see Figure 4.57(b); they may be attributed to the brittleness of the superconducting ceramic and to the excessive shock pressure developed during the detonation of the explosive; the peak shock pressure was about 10 GPa, see Table 4.4.

After firing, as the detonation wave proceeds along the cladding direction, transition from laminar to turbulent flow occurred, resulting in the change of the initial straight (waveless) interface to a wavy von Kárman pattern. Optical micrographs revealed such a wavy interface, see Figure 4.58(a), whereas an enlarged view of the vortex between the flyer plate and the base plate is presented in Figure 4.58(b). This is one of the main characteristic phenomena of explosive cladding [47,48], and it may be also explained by the dynamic plasticity mechanism outlined in Reference [49].

From microhardness measurements in the same region, close to the weld zone, it is evident that the average microhardness of the flyer plate is greater (110 HV) than that of the parent plate (90 HV), probably because of the high detonation velocity explosive used. For both cases, their measured average microhardness is greater than the initial microhardness of the silver plate, before cladding (approximately 80–85 HV). The microhardness increase in both the flyer and the parent plates may be attributed to the intense stress waves, associated with the high-velocity impact. Besides the overall increase in hardness, localized shock-induced hardening occurred at the interface and also in the vicinity of the interface, between the parent plate and the steel anvil, probably because of the superposition of the transmitted and reflected stress waves at the interface, see Figure 4.59.

Scanning electron microscopic observations revealed a very fine microstructure of the explosively consolidated material, in comparison with the initial screen printed superconducting powder of the thick film; compare Figures 4.25 and 4.60. An abrupt reduction of the average grain size occurred because of the intense grain fragmentation that resulted from shock wave powder consolidation. Consequently, many grain boundaries, resulting in a decrease of the microporosity, and a large number of lattice imperfections were formed; they can trap the fluxoids, improving, therefore, the supercurrent conducting ability (Type II superconductivity) of the superconducting film.

X-ray diffraction patterns, obtained by applying the X-ray diffraction technique on the superconducting thick film (screen printed on silver substrate)

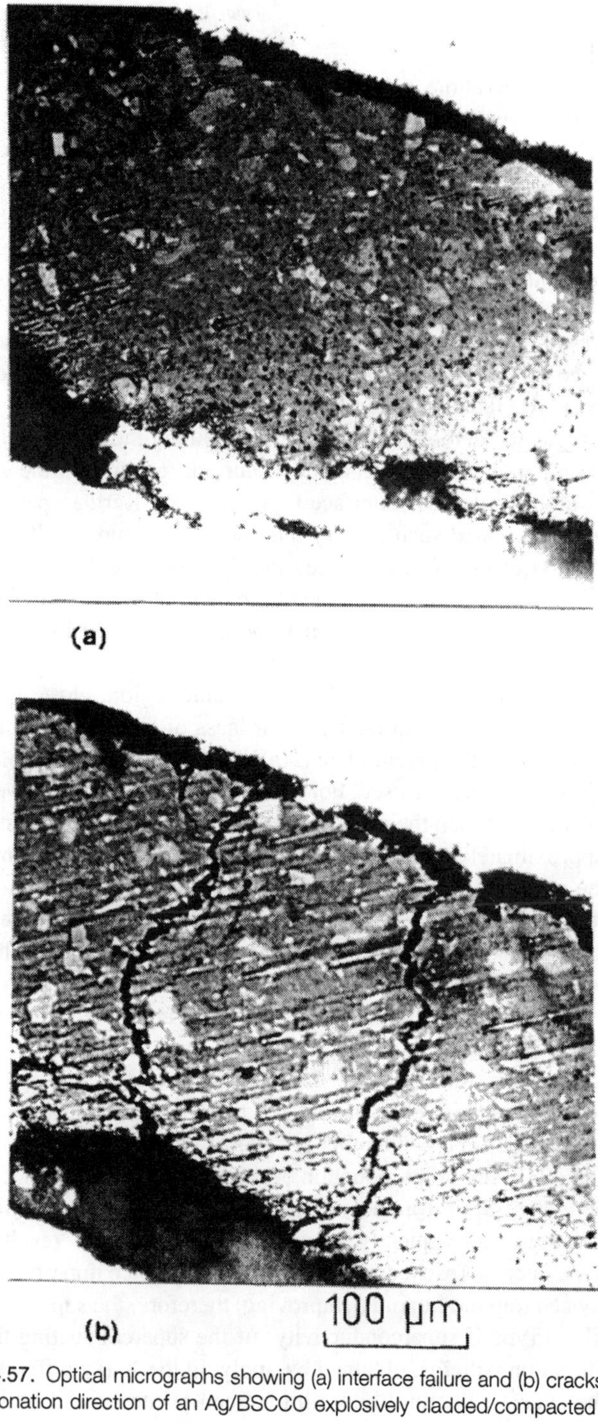

FIGURE 4.57. Optical micrographs showing (a) interface failure and (b) cracks perpendicular to detonation direction of an Ag/BSCCO explosively cladded/compacted plate.

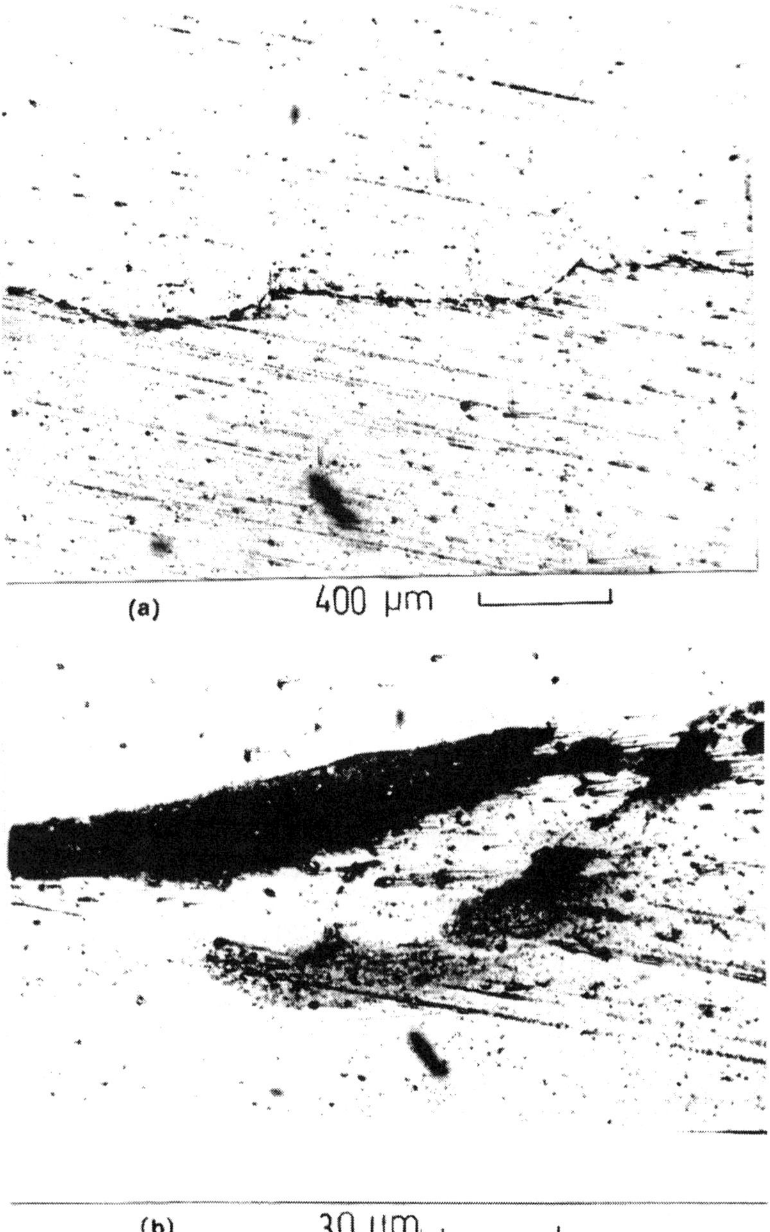

FIGURE 4.58. Optical micrographs showing (a) a wavy interface and (b) an isolated vortex of an Ag/BSCCO explosively cladded/compacted plate.

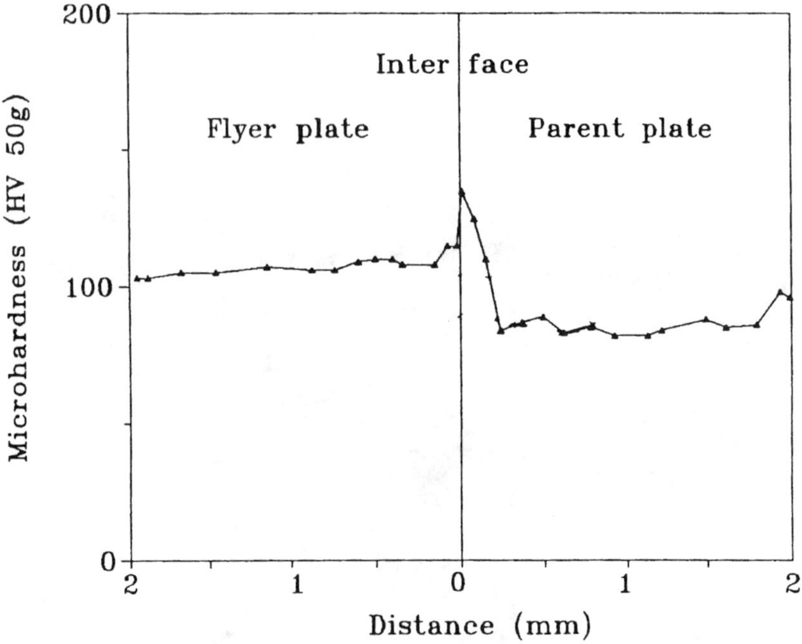

FIGURE 4.59. Microhardness variation as a function of distance from the cladding interface for an Ag/BSCCO explosively cladded/compacted plate.

and on the shock-induced consolidated superconducting layer, revealed a multiphase microstructure in both cases, see Figures 4.26(a) and 4.61(a). For the screen-printed superconducting material, the three main phases of the Bi-Pb-Sr-Ca-Cu-O compound, namely the 2201, the 2212 and the 2223 phases, were detected; their mass was 9%, 12% and 78% w/w of the total mass, respectively. Note also that a small quantity, less than 1% w/w, of an impurity corresponding to the Ca_2PbO_4 phase was also detected. For the explosively compacted plate, an increase in width of diffraction peaks was observed, indicating a higher disorder or incompleteness in crystallinity and also a significant reduction of the grain size due to the high shock pressures developed during the explosive compaction of this film; compare also with similar observations reported in Reference [16]. The mass of the detected phases 2201, 2212, 2223 was 11%, 21% and 67% w/w of the total mass, respectively. Therefore, it may be concluded that partial transformation of the 2223 phase, about 11%, into 2212 and 2201 phases occurred, probably because of the excessive heating obtained during the shock compaction.

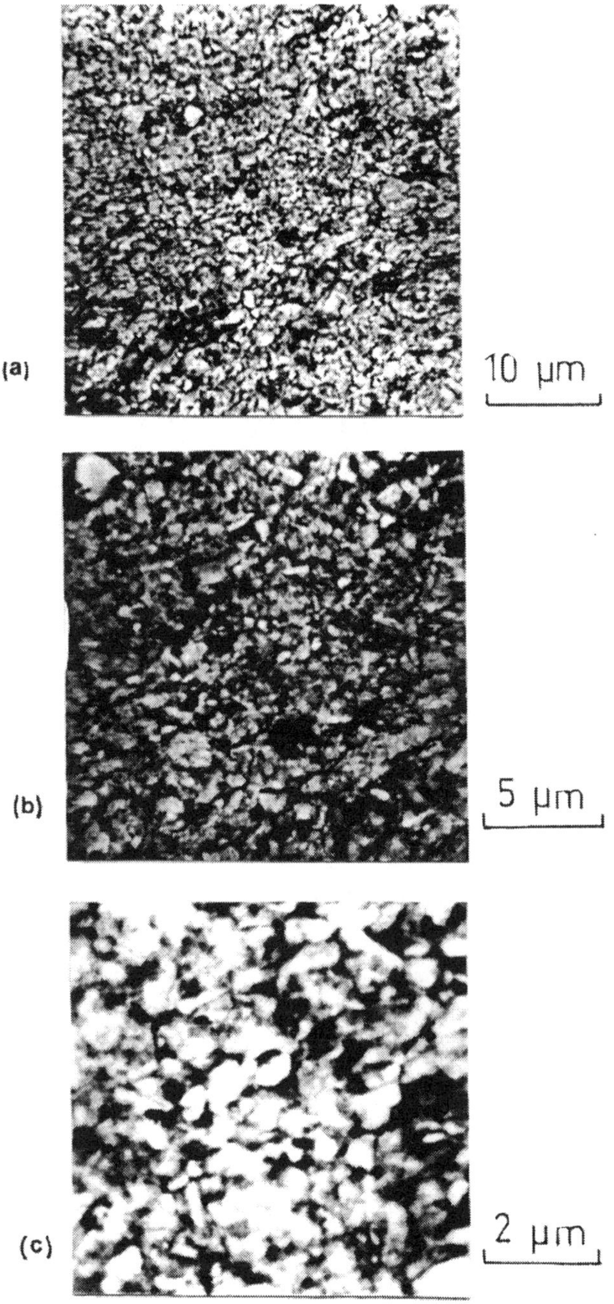

FIGURE 4.60. Scanning electron micrographs showing the microstructure of the compacted BSCCO ceramic.

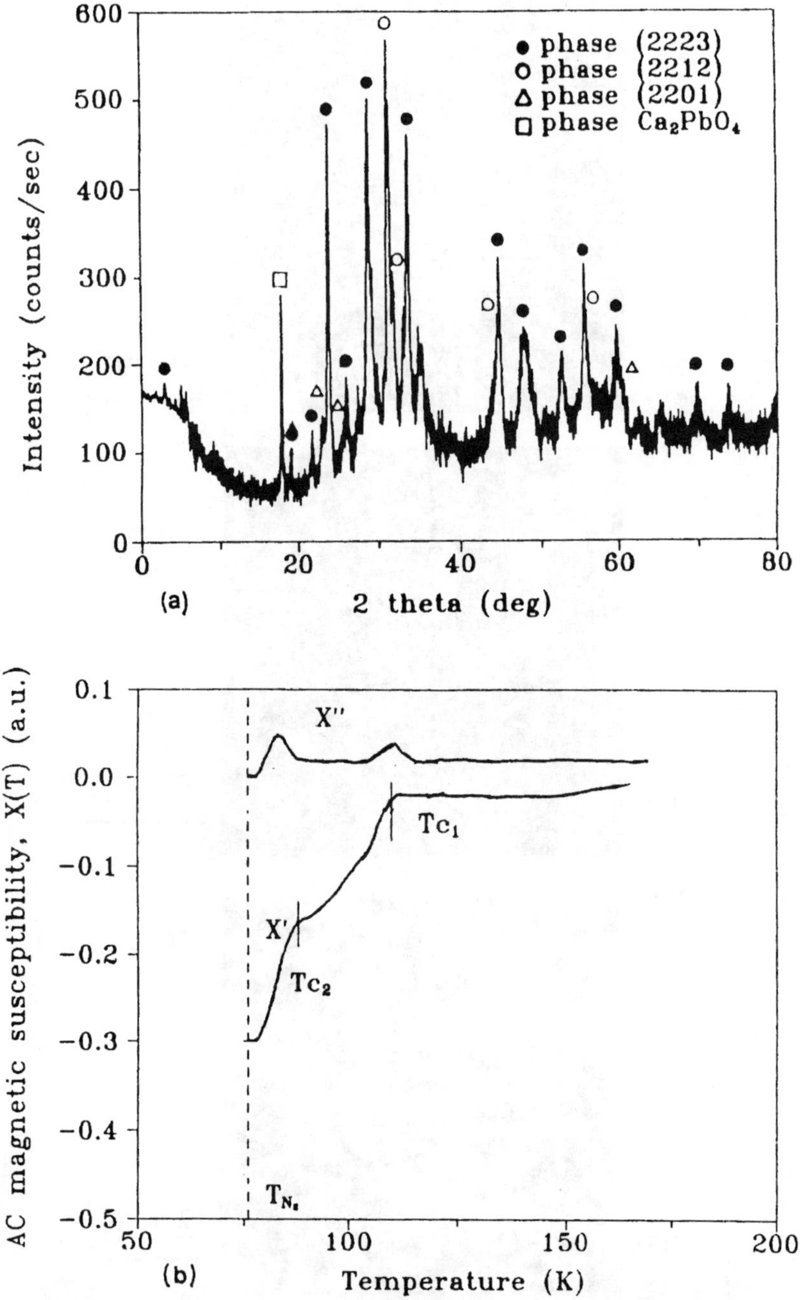

FIGURE 4.61. (a) XRD spectrum corresponding to the compacted BSCCO ceramic. (b) The ac-susceptibility curves of the compacted BSCCO ceramic as a function of temperature.

The ac-magnetic susceptibility measurements, performed on the initial superconducting thick film, revealed two transition points at:

- $T_{c1,onset} = 108$ K, with $\Delta T_{c1} = T_{c1,onset} - T_{c1,offset} = 12$ K, resulted from the lead doped 2223 phase.
- $T_{c2,onset} = 86$ K, with $\Delta T_{c2} = T_{c2,onset} - T_{c2,offset} = 8$ K, resulted from the lead doped 2212 phase, see the diamagnetic susceptibility curve, χ' in Figure 4.26(b).

For the ac-magnetic susceptibility of the explosively compacted sample in relation to the temperature, from Figure 4.61(b) it is evident that two relatively broader transition regions were apparent, namely at:

- $T_{c1,onset} = 110$ K, with $\Delta T_{c1} = 20$ K, resulted from the lead doped 2223 phase.
- $T_{c2,onset} = 87$ K, with $\Delta T_{c2} = 10$ K, resulted from the lead doped 2212 phase.

Although the expected transitions, in the case of the explosively consolidated ceramic, occurred at slightly higher temperatures, compared with the transitions of the initial pasted material, the superconductivity of the explosively compacted sample was reduced, as it is evident from the transition width and the intensity of the Meissner signals; compare Figures 4.26(b) and 4.61(b). This reduction of superconductivity may be attributed to:

- The crystal imperfections formed after explosive compaction of the powder.
- The additional oxygen loss and, subsequently, the formation of oxygen vacancies is due to the high-pressure shock loading, see References [50]–[52]. Note that because of the lattice defects and oxygen nonstoichiometry, the carriers in the Cu-O layers or Bi-O layers weaken the metallic property in the normal (nonsuperconducting) state and, therefore, reduce superconductivity. On the other hand, the band structure near the Fermi level may be altered, affecting, therefore, the superconductivity [51].
- The above-mentioned partial transformation of the high-T_c 2223 phase to the low-T_c 2212 phase and to the semiconducting 2201 phase.

4.6 REFERENCES

1. Mamalis A. G., Szalay A., Pantelis D. I., Pantazopoulos G., Kotsis I. and Enisz M. (1996), "Fabrication of multi-layered steel/superconductive ceramic (Y-Ba-K-Cu-O)/silver rods by explosive compaction and extrusion," *J. Mater. Proc. Technol.* 57, 155.

2. Mamalis A. G., Szalay A., Pantelis D. I. and Pantazopoulos G. (1996), "Net shape manufacturing of silver-sheathed high-T_c superconductive ceramic (Y-Ba-K-Cu-O) strip by explosive compaction/cladding and rolling," *J. Mater. Proc. Technol.* 57, 112.

3. Murr L. E. (1991), "Explosive processing of bulk superconductors," *Mater. Manuf. Proc.* 6, 1.

4. Kondo K. I., Soga S. and Sawaoka A. (1985), "Shock compaction of silicon carbide powder," *J. Mater. Sci.* 20, 1033.

5. Wang S. L., Meyers M. A. and Graham R. A. (1986), "Shock-consolidation of IN-100 nickel-base superalloy powder," in *Shock Waves in Condensed Matter.* Y. M. Gupta, ed. Plenum, New York, 731.

6. Prümmer R. (1988), "Explosive compaction of metallic glass powders," *Mater. Sci. Eng.* 89.

7. Blazynski T. Z. (1986), "Explosive compaction of ceramic and polymeric powdered," in *Metallurgical Applications of Shock-Wave and High-Strain-Rate Phenomena.* L. E. Murr and K. P. Staudhammer, eds. Marcel Dekker, New York, 189.

8. Blazynski T. Z. (1993), "Explosively consolidated PVC-alumina powder mixtures," *J. Mater. Proc. Technol.* 39, 389.

9. Bergmann G. and Bailey N. F. (1987), *High Pressure Explosive Processing of Ceramics.* R. A. Graham and A. B. Sawaoka, eds. Switzerland, 165.

10. Saito S. and Sawaoka A. B. (1979), *Proc. 7th AIRAPT Conf. on High-Pressure Energy Science and Technology,* Leckvesan, France, 541.

11. Prümmer R. (1983), *Explosive Welding, Forming and Compaction.* T. Z. Blazynski, ed. Applied Science, London.

12. Meyers M. A. (1994), *Dynamic Behavior of Materials.* John Wiley & Sons, New York.

13. Meyers M. A. and Wang S. L. (1988), "An improved method for shock consolidation of powders," *Acta Metall.* 36, 925.

14. Iqbal Z., Rao K. V., Puzniac R., Thadhani N. N. and Ramakrishna B. L. (1990), "Enhanced intra- and inter-grain critical currents in shock wave processed $YBa_2Cu_3O_7$," *Proc. ICME 90 Topical Conf. on High-Temperature Superconductors, Materials Aspects,* Garmisch-Paterkirchen, Germany.

15. Edwards M. R., Rogers K. D., Spencer J. W. C., Denton I. E. and Briggs A. (1991), "Preliminary observations of the explosive compaction of superconducting oxides in the Bi-Sr-Ca-Cu-O system," *J. Mater. Sci. Lett.* 10, 181.

16. Yoshimoto M., Koinuma H., Yamamoto H. and Sawaoka A. B. (1990), "Shock processing of Bi-Sr-Ca-Cu-O and Tl-Ba-Ca-Cu-O superconductors," *J. Am. Ceram. Soc.* 73, 1791.

17. Murr L. E., Niu C. S. and Pradhan-Advani M. (1991), "Effect of shock pressure on superconductivity in explosively fabricated Y-Ba-Cu-O/metal matrix composites," *Phys. Stat. Sol. A* 123, 507.

18. Mamalis A. G., Gioftsidis G. N. and Prohászka J. (1990), "Manufacturing of thin rectangular plates by explosive compaction of copper powder," *Proc. I. Mech. E. B* 204, 237.

19. Crossland B. (1982), *Explosive Welding of Metals and Its Applications.* Clarenton Press, Oxford.

20. Prümmer R. (1987), *Explosivvertichtung Pulvriger Substanzen.* Springer, Berlin.

21. Mamalis A. G., Gioftsidis G. N., Szalay A. and Boday O. (1989), "The shock wave compaction of high temperature superconducting powders," *CIRP Ann.* 38, 297.

22. Mamalis A. G., Szalay A., Pantelis D. I. and Pantazopoulos G. (1995), "Net shape manufacturing of metal/superconductive ceramic/metal rods by explosive compaction and warm extrusion," *Proc. EXPLOMET' 95, Int. Conf. Metallurgical and Materials Applications of Shock-Wave and High-Strain-Rate Phenomena,* El Paso, Texas. L. E. Murr, K. P. Staudhammer and M. A. Meyers, eds. Elsevier, New York, 747.

23. Wang S. L., Meyers M. A. and Szecket V. A. (1988), "Warm shock consolidation of IN 718 powder," *J. Mater. Sci.* 23, 1786.

24. Ferreira A., Meyers M. A., Chang S. N., Thadhani N. N. and Kough J. R. (1991), *Metall. Trans. A* 22, 685.

25. Mamalis A. G., Szalay A., Göbl N., Vajda I. and Raveau B. (1998), "Near net-shape manufacturing of metal sheathed superconductors by high energy rate forming techniques," *Mater. Sci. Eng. B* 53, 119.

26. Williams D. J. (1981), "Compaction of metal powders using high voltage discharges," Ph.D. thesis, University of Cambridge.

27. Hugoniot H. (1889), *J. de l'Ecole Polytechnique* 58, 3.

28. Bridgman P. W. (1949), *Proc. Am. Acad. Arts Sci.* 77, 189.

29. Yeremin E. N. (1978), *Chemical Thermodynamics.* Mir Publishers, Moscow.

30. Schwarz R. B., Kasiraj P., Vreeland T. and Ahrens T. J. (1984), "A theory for the shock-wave consolidation of powders," *Acta Metall.* 32, 1243.

31. Carroll M. M. and Holt A. C. (1972), "Static and dynamic pore-collapse relations for ductile porous materials," *J. Appl. Phys.* 43, 1626.

32. Carroll M. M., Kim K. T. and Nesterenko V. F. (1986), "The effect of temperature on viscoplastic pore collapse," *J. Appl. Phys.* 59, 1962.

33. Nesterenko V. F. (1992), *High Rate Deformation of Heterogeneous Materials.* Nauka, Novosibirsk, Russia.

34. Verwerft M. G. M., Dijken D. K., de Hosson J. T. M. and van der Steen A. C. (1994), "Different types of dislocations in $YBa_2Cu_3O_{7-x}$," *Phys. Rev. B* 50, 3271.

35. Dijken D. K. (1994), "Dynamic and isostatic densification of powder materials," Ph.D. thesis, University of Groningen, The Netherlands.

36. Gurney R. K. (1943), "The initial velocities of fragments from bombs, shells and grenades," Report No. 405, Ballistics Research Laboratory (BRL), Aberdeen, MD.

37. Singh J. B., Joo J., Singh D., Warzynski T. and Poeppel R. B. (1993), "Effects of silver additions on thermal shock and delayed failure of $YBa_2Cu_3O_{7-\delta}$," *J. Mater. Res.* 8, 1226.

38. Besenyei E., Katona G., Arato P. and Kele A. (1989), "The effect of heat treatment on the superconducting properties of Bi-Sr-Ca-Cu thick films on alumina substrates," *Supercond. Sci. Technol.* 2, 220.

39. Pissas M., Nicolaides G. K., Psycharis V. and Niarchos D. (1992), "Quantitative analysis and studies of the transformation $Bi_2Sr_2CaCu_2O_{8+x}$ to $Bi_2Sr_2Ca_2Cu_3O_{10+x}$ using Rietveld analysis and ac-susceptibility," *Physica C* 196, 157.

40. Bean C. D. (1964), "Magnetisation of high field superconductors," *Rev. Mod. Phys.* 36, 31.

41. Yoffé E. H. (1951), "The moving Griffith crack," *Phil. Mag.* 42, 739.

42. Mamalis A. G., Kotsis I., Pantazopoulos G., Enisz M., Szalay A. and Manolakos D. E. (1997), "Effect of heat-treatment on explosively compacted (Y-Ba-K-Cu-O) superconductive powders," *Physica C* 280, 289.

43. Kotsis I., Enisz M., Oravetz D. and Szalay A. (1995), "Effect of porosity on properties of explosively compacted high-T_c superconductors," *Hungarian J. Ind. Chem. Vészprem* 23, 69.

44. Enisz M., Kotsis I., (1990), "Relation between the physical properties and microstructure of $YBa_2Cu_3O_x$-based ceramic superconductors containing K^+ and Na^+ ions," *Proc. 7th CIMTEC Satelite Symposium,* Trieste, Italy, 135.

45. Malik S., Mohammad M., Khan A. Y. and Subhani M. S. (1993), "Superconductivity in Y-Ba-Na-Cu-O," *J. Mater. Sci. Lett.* 12, 814.

46. Mamalis A. G., Szalay A., Pantelis D. I. and Pantazopoulos G. (1995), "Fabrication of thick layered superconductive ceramic (Bi-Pb-Sr-Ca-Cu-O)/metal composite strips by explosive cladding and rolling," *J. Mater. Proc. Technol.* 51, 251.

47. Jaramilo D., Szecket V. A. and Inal O. T. (1987), "On the transition from a waveless to a wavy interface," *Mater. Sci. Eng.* 91, 217.

48. Linse V. D. and Lalwaney N. S. (1984), "Explosive welding," *J. Met.* 35, 62.

49. Szecket A., Vigueras D. J. and Inal O. T. (1986), "The cyclic pressure distribution of explosively welded interfaces," *Proc. Metallurgical Applications of Shock-Wave and High-Strain-Rate Phenomena,* Marcel Dekker, New York, 887.

50. Matizen E. V., Deribas A. A., Nesterenko V. F., Pershin S. A., Berzverkhii P. P., Voronin A. N., Yefermova R. I. and Starikov M. A. (1989), "Effect of high dynamic pressures on the properties of high-T_c ceramics," *Int. J. Mod. Phys. B* 3, 97.

51. Nanlin W., Mingqiu T., Jinsong W., Zhuan X., Jian S., Shengdi X. and Qirui Z. (1991), "Effect of heat treatment on the superconductivity of 110 K single phase in the Bi(Pb)-Sr-Ca-Cu-O system," *J. Mater. Sci. Lett.* 10, 214.

52. Niu H., Fukushima N. and Ando K. (1988), "Effect of oxygen content and Sr/Ca ratio on superconducting properties in $Bi_2Sr_{2-x}Ca_{1+x}Cu_2O_{8+\delta}$," *Jap. J. Appl. Phys. Lett.* 27, 1442.

CHAPTER 5

Fabrication of Bulk HTS

5.1 NOTATION

a = lever arm of roll force
A_o = initial cross-sectional area of extruded billet
A_f = final cross-sectional area of extruded billet
A_i = cross-sectional area of ith layer of multilayer body
b = width of strip
b_1 = width of strip at entry of roll gap
b_2 = width of strip at exit of roll gap
b_m = mean width of rolled strip $[(b_1 + b_2)/2]$
c = crack length
D_o = initial diameter of extruded billet
F = applied load
F_R = roll force
h = thickness of composite strip
h_1 = thickness of composite strip at entry of roll gap
h_2 = thickness of composite strip at exit of roll gap
h_i = average thickness of ith layer of multilayer strip
J_c = critical current density
K_{Ic} = stress intensity factor
L = length of extruded billet in the container
p = normal pressure
P = punch load
R = extrusion ratio
R' = deformed roll radius
r_T = total reduction in the cross-sectional area of component
T_c = critical transition temperature

T_m = melting point
T_R = roll torque
x = horizontal stress due to conventional extrusion
x_R = horizontal stress due to conventional rolling
Y = yield stress
Y^* = effective yield stress of multilayer component
$\bar{Y}$ = mean effective yield stress
y_i = transmitted shear stress at the interface between layers
β = h_2/h_1
Δ = deformation factor
Δh = $h_1 - h_2$
ΔT_c = transition temperature width
δ = punch travel
ϵ = true strain
μ = coefficient offriction
σ_{1i} = principal horizontal stress at ith layer of multilayer component
σ_3 = principal vertical stress
φ = semiapical angle of extrusion die

5.2 GENERAL

Superconducting high-power electrical and electromagnetic applications require components that exhibit not only special physical properties but also have specific geometries. Rectangular silver-sheathed superconducting plates, possessing a high-critical current density, J_c, are designed and fabricated as necessary elements of high-power electrical machines and levitated bearings. Cylindrical shapes are required for high-current electrical power transmission by using silver-clad multicore superconducting cables. The same shaped components, in a shorter length, are also used as high-T_c conductors in electrical generators and, after appropriate twisting, as high-field superconducting coils. The selection of the appropriate manufacturing process for the fabrication of the required bulk ceramic material encounters for the soundness, i.e., mechanical integrity, strength, structural stability, etc., and, therefore, the workability, i.e., current conduction, magnetic field induction, etc., of the final component.

Among the manufacturing processes used in the superconductor industry is the *powder-in-tube* technique (*PIT*). Industrial-scale production of transmission cables and tapes is realized by using the PIT process. Conventional deformation processes, such as rolling and extrusion, are also used for the fabrication of bulk superconducting components, namely metal-sheathed and precompacted ceramic core plates and rods. Compression-like deformation processing improves the plastic flow and, therefore, the formability of the ceramic core material, depressing the formation of various defects, such as bursts and microcracks. The selection of the optimum intermediate, or final heat-treating conditions helps also in the re-establishment of the necessary superconducting properties of the component.

5.3 PLASTIC DEFORMATION: THEORETICAL MODELING

5.3.1 Workability

The production of bulk superconducting components must take into account a large number of working parameters, related to the various stages of fabrication, see the flowchart in Figure 1.4, i.e., powder synthesis methods, deformation processing and heat treatment parameters. In this section, the effect of deformation processing on the macro- and microfailures, microstructure, morphology, uniformity, density and superconducting properties of the ceramic oxides are discussed.

A classification of the working processes may be made by reference to the stress system on an element in a particular process. Changes in shape are due to the shear stress components, whereas all-round hydrostatic stress helps determine ductility; compressive hydrostatic stress improves ductility by suppressing void and fracture formation, which is promoted by tensile hydrostatic stress. The different deformation processes are related to the indentation or open-die forging of a block by opposed flat rigid dies as shown in Figure 5.1, see also References [1] and [2].

PLASTIC DEFORMATION AND FRACTURE

Among other factors governing surface and internal failures are the ductility of the material, temperature and stress field (and the magnitude of the local hydrostatic pressure) around the position.

The presence of cracks in a superconducting core of a rod or a plate produced by deformation processing results in the deterioration of the superconducting properties. It is, therefore, necessary to reduce the crack formation during processing or to close these cracks after postprocessing heat treatment, by applying compressive stress fields. It has been reported that, in many cases, the soundness and, subsequently, the superconducting performance of tapes that are deformed by forging between rigid dies for the second and subsequent deformation stages is probably better than that of equivalent tapes deformed by rolling [3]. This may be attributed to the amount of compression undergone in the first process, which is greater than that exhibited when rolling the strips with small reductions [1].

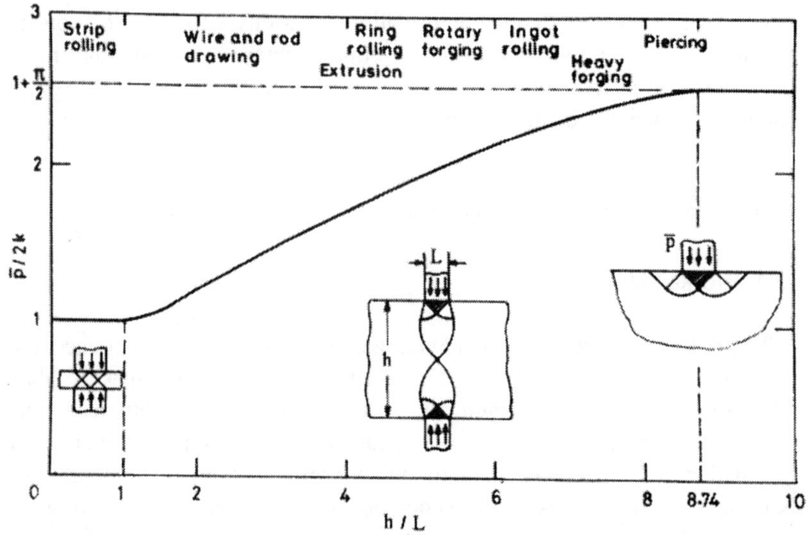

FIGURE 5.1. Classification of forming processes. Evolution of $\bar{p}/2k$ as a function of deformation factor Δ (= h/L).

Failures occurring during compression-like working processes, such as rolling, forging and extrusion, are predominantly caused by so-called "secondary" tensile stresses. Fracturing at the edges, i.e., edge cracking, due to limited ductility, variation of the stresses along the width of the rolled material and uneven deformation at the edges, is profound when rolling relatively thick plates with small reduction per pass. Transverse cracks may propagate from the edges of a strip, if too large reduction is attempted without intermediate heat treatment, whereas fir cracking, i.e., the development of transverse cracks, during heat treatment after strip rolling, is also expected for certain rolling conditions [2].

Longitudinal cracking along the direction of plastic flow due to the presence of secondary tensions is developed when upsetting ceramic plates and rods. Cracks due to the tangential velocity discontinuities, i.e., zones of localized plastic deformation in the direction of maximum shear stress, may arise in forging of insufficiently ductile materials [2].

Cracking due to hot or cold shortness, in the form of repetitive peripheral cracks, due to secondary tensile stress fields developed in the region of the die, may occur during the extrusion of rods of brittle materials. Low-ductility extruded materials are prone to radial and circumferential cracking, related to strain inhomogeneity. Internal cracks, which appear to follow extrusion flow lines, and central bursts, i.e., internal arrow-shaped defects, are occasionally encountered in extrusion and wire-drawing; upper bound approaches and fracture criteria have been proposed to formulate these defects, see References [1] and [2].

The fracture toughness of the ceramic core, expressed by means of the stress intensity factor K_{Ic}, constitutes a measure of the crack propagation sensitivity and, therefore, it is an important characteristic of the soundness of the produced component; it can be estimated as a function of the $F/c^{3/2}$, where F is the applied load and c the crack length.

TEXTURE AND MICROSTRUCTURE

The deformation processing of a ceramic core component is likely to have a significant effect on the crystallographic texture and the reaction kinetics during thermal treatment. The texture of the ceramic core leads to beneficial transport properties and high-current densities. The texture development is based on polycrystalline plasticity and rigid particle rotation models. Frictional gliding, cracking and rotating of brittle individual grains are the basic microscopic processes, involved in the microstructure development during deformation processing. The slip-plane, e.g., of the crystals of the 2223 phase in the BSCCO system, is the c-plane (001). When shear deformation occurs, crystals rotate in the orientation in which the slip-plane is parallel to the shear-plane. In strip rolling, shear deformation parallel to the rolling plane occurs more easily when the deformation zone geometry factor, Δ is small, around unity and the slip-lines penetrate the whole section of the strip, see Figure 5.1. The increase of the number of passes promotes texturing, whereas an optimum reduction value may be defined, where maximum texturing is achieved.

The microstructural and morphological development during forming can be evaluated by optical metallography, SEM and by X-ray pole-figure analysis. A typical example of texture development is found in BSCCO tapes produced by the PIT technique; 2223 phase grain alignment during working forms a highly oriented core structure, see Figure 5.2(a).

DENSITY AND UNIFORMITY

The density of the ceramic core is greatly affected from the imposed plastic work, i.e., by increasing the reduction to a certain critical value. Above this critical level, catastrophic cracks, delaminations and ceramic core inhomogeneities may be observed. Compaction characteristics depend on powder microstructure, particle shape, size distribution, morphology and initial powder density, along with the intrinsic mechanical properties of the powder constituents. These properties affect the three general stages of powder consolidation during deformation processing: restacking, deformation and fracture. It may be noted that, the BSCCO ceramic system exhibits greater densification than the YBCO ceramic system; restacking via rotation and gliding is

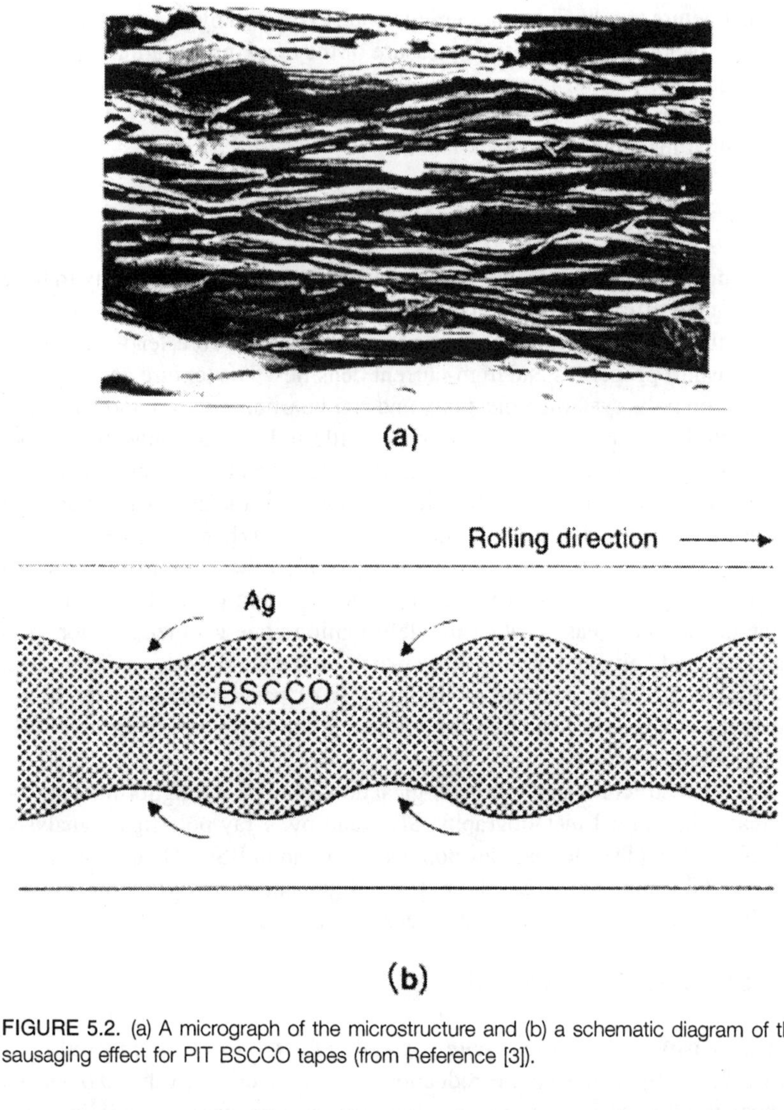

FIGURE 5.2. (a) A micrograph of the microstructure and (b) a schematic diagram of the sausaging effect for PIT BSCCO tapes (from Reference [3]).

abetted by its platelet morphology, whereas limited plasticity is realized through the {001}<100> slip-system in the 2212 phase.

The characterization of the ceramic core morphology is based on the control of uniformity and connectivity among grains. In rolling of superconducting tapes, large reduction passes and small diameter rolls (small length of contact between rolls and the workpiece), i.e., the deformation factor $\Delta > 1$, (see Figure 5.1) lead to nonuniform core, resulting very often in discontinuities and gaps through the whole section of the strip, because deformation is confined more near the contact surface not completely penetrating the whole cross section. The inner core discontinuities, formed during processing, constitute the sausaging effect, see Figure 5.2(b). Sausaging, caused by inhomogeneous plastic deformation, results in localized reduction of the effective supercurrent conduction cross section, leading, therefore, in the subsequent decrease of the critical current density, J_c.

Another microstructural parameter affected by the progress of deformation is the grain connectivity. The connectivity of grains of the oxide core is sensitive to the pressure applied during shaping. By increasing the reduction, the grain connectivity is enhanced, but it is reduced when increasing the number of passes [4].

Depending on the materials processing conditions, weak grain-links, i.e., porosity, secondary phase formation, grain boundary microcracks and lattice disorder in the near-interparticle area, may be formed. They greatly affect the intergrain current conduction, reducing, therefore, the transport capability of bulk polycrystalline superconductors.

CRITICAL CURRENT DENSITY

As mentioned above, changes of texturing and grain connectivity due to deformation processing greatly affect the critical current density of the superconductor, as qualitatively illustrated in Figure 5.3. The increase of both texturing and connectivity enhances the critical current density. Also presented in Figure 5.3 is the resultant change of J_c, as a function of reduction and number of passes, by taking into consideration the contributions of texturing and connectivity to the critical current density; see also Reference [4].

5.3.2 Forming of Multilayer Composites: a Theoretical Approach

The forming of multilayered 3D and axisymmetric HTS composites has been thoroughly examined by the authors [5]–[12]. The direct multiplepass rolling and extrusion processes have been applied to further shaping metal-HTS composite plates and billets, fabricated by the explosive compaction/cladding technique, described in detail in Chapter 4, into metal-sheathed superconducting strips, rods and wires.

ROLLING

A theoretical estimation of the overall strength of cladded multilayer plates is important because of the industrial need for highly functional composite sheets. Based on the concept of a "multilayer" body and on the "sandwich"

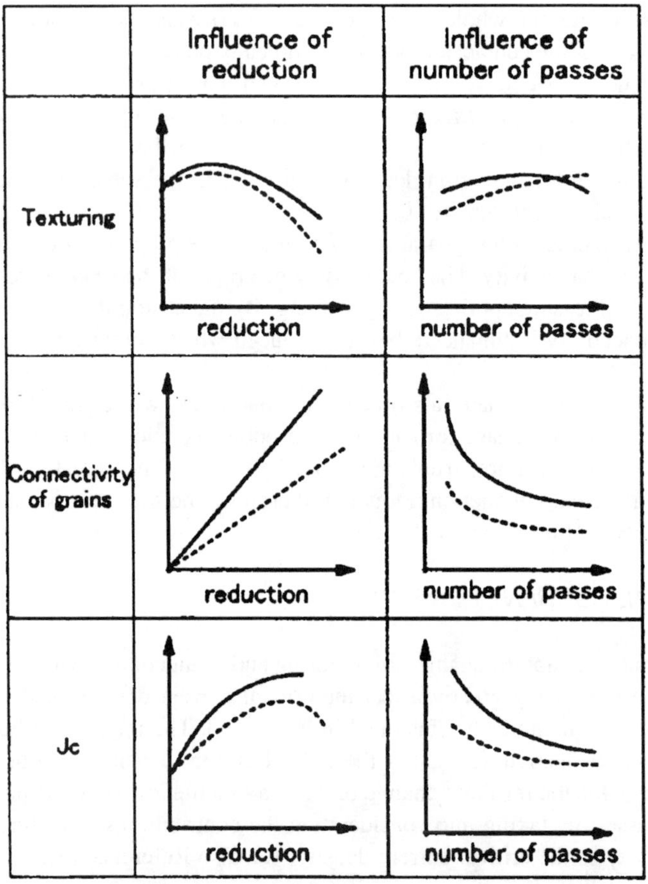

FIGURE 5.3. Schematic illustration of variations of texturing, connectivity and critical current density with reduction and number of passes, when rolling metal sheathed BSCCO tapes. (solid line) Rolling with large diameter rolls; (dashed line) rolling with small diameter rolls (from Reference [4]).

rolling technique and by considering the cladded-rolled strip as a two- or three-layer body (see Figure 5.4) a theory of the rolling of cladded bimetallic and trimetallic strips was presented in References [13] and [14]; a modification of this theory to account for the metal/ceramic/metal and the metal/metal cladded plates configuration in Figure 4.22 and also References [5] and [6] is outlined below.

In the first case, the outer layers (a) and (b), [see Figure 5.4(a)] represent the layers of the two initial metals of the parent and the plyer plate, whereas the intermediate layer (c) represents the ceramic core. Note that the interme-

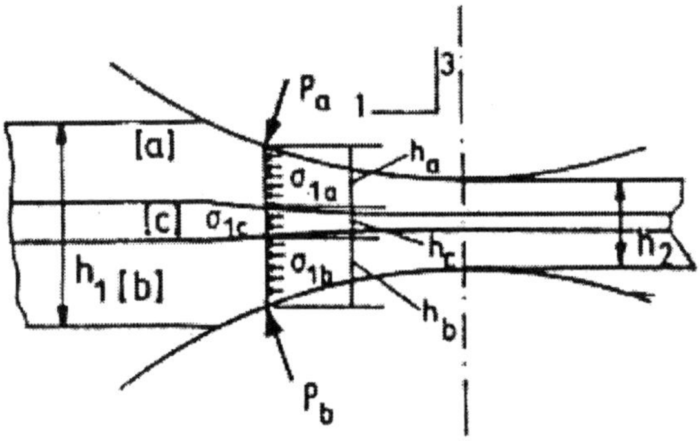

(i)

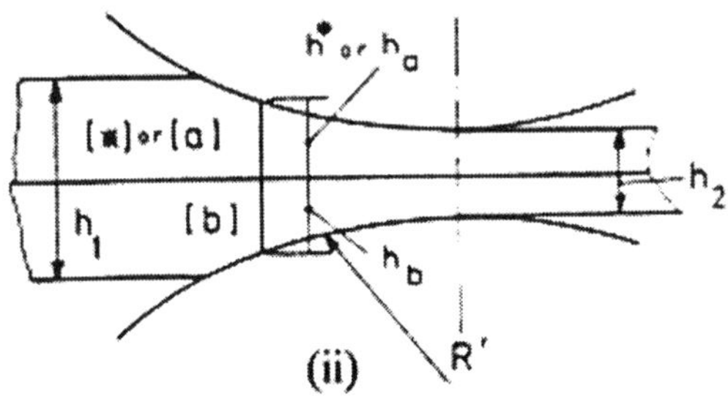

(ii)

FIGURE 5.4. (a) Stress field in the deformation zone during rolling of cladded three-layer composite strips. (b) Modelling of materials when rolling cladded two-layer composite strips.

diate layers, representing the very narrow transition zones at the metal/ceramic interfaces, are very small, seldom exceeding a few microns as revealed from metallographic studies, and, therefore, they may be ignored.

For plane-strain conditions and with the simplified assumptions made, assuming that the materials may be considered as rigid-perfectly plastic, the deformed arc of contact (radius R') is circular, and the coefficient of friction between the outer layers and the rolls is constant over the arc of contact, and, furthermore, because the outer softer layers are rolled to a slightly greater elongation than the inner harder core frictional forces are induced between the layers [see Figure 5.4(a)] yielding in the three layers, when the Tresca criterion is considered, results in:

$$\begin{aligned} \sigma_3 - \sigma_{1a} &= Y_a \\ \sigma_3 - \sigma_{1c} &= Y_c \\ \sigma_3 - \sigma_{1b} &= Y_b \end{aligned} \tag{5.1}$$

It is taken that the normal pressures p_a and p_b, [see Figure 5.4(a)] are identical and for the plane-strain rolling conditions considered they equal the vertical stress, i.e., $p_a = p_b = \sigma_3$ and that the horizontal stresses σ_{1i} ($i = a, b, c$) are distributed uniformly over the corresponding vertical sections is summing the stress developed in the relevant layer due to conventional rolling, x_R and the transmitted shear stress at the relevant interface, y_i,

$$\begin{aligned} \sigma_{1a} &= x_R + y_a \\ \sigma_{1c} &= x_R - y_c \\ \sigma_{1b} &= x_R + y_b \end{aligned} \tag{5.2}$$

Equilibrium of the additional stresses due to restraint at the interfaces results in

$$y_a h_a + y_b h_b = y_c h_c \tag{5.3}$$

The effective yield stress Y^* for "sandwich" rolling may be denoted as [12]

$$Y^* = \sigma_3 - x_R \tag{5.4}$$

and by combining Equations (5.1) and (5.2) and inserting in Equation (5.4), the effective yield stress, Y^* can be estimated as

$$Y^* = \frac{Y_a h_a + Y_b h_b + Y_c h_c}{h_a + h_b + h_c} \tag{5.5}$$

When two-layer strips are considered [see Figure 5.4(b)] with the same above-mentioned assumptions made, and taking into account that the transition zone at the interface, also in metals, seldom exceeds 100 μm [13], Equation (5.5) is simplified to

$$Y^* = \frac{Y_a h_a + Y_b h_b}{h_a + h_b} \tag{5.6}$$

It must be noted that h_a, h_b and h_c refer to the average thickness of the corresponding layer, i.e., $h_i = (h_{ientry} + h_{iexit})/2$ and, therefore, the effective yield stress, Y^*, obtained from Equations (5.5) and (5.6), may be designated as the mean effective yield stress, $\bar{Y}$ to the rolling conditions examined.

For plane-strain conditions, i.e., if spread is negligible, an estimation of the roll force, F_R may be obtained from the modified Ekelund's formula [12],

$$F_R = \mathrm{b}\bar{Y}\sqrt{R' \cdot \Delta h}\left(1 + \frac{1.6\mu\sqrt{R' \cdot \Delta h} - 1.2\Delta h}{h_1 + h_2}\right) \quad (5.7)$$

where h_1, h_2 refer to the overall thickness of the sandwich plate before and after rolling pass, respectively, $\Delta h = h_1 - h_2$, R' the deformed roll radius and μ the coefficient of friction between the rolls and the outer layers of the plate; $\bar{Y}$ is the mean yield stress of the material estimated from Equations (5.5) and (5.6) for the three- and two-layer composite, respectively.

Roll-flattening is taken into account by using the deformed roll radius R', which is obtained by using Hitchock's formula and a graphical procedure, see also Reference [15]. For slabs and for thickness reduction greater than 50% some spread is expected and the active width of the strip, b in Equation (5.7) may be substituted by a mean thickness $b_m = (b_1 + b_2)/2$ to account for the spread.

The torque required to drive both rolls is calculated as follows:

$$T_R = 2F_R\alpha \quad (5.8)$$

where, following Wusatowski [15], the lever arm of roll force, α is given as

$$\alpha = 0.5\sqrt{R' \cdot \Delta h}\,\frac{\beta}{1 - \beta}\ln\left(\frac{1}{\beta}\right) \quad (5.9)$$

EXTRUSION

The configuration of the cladding/compaction of the axisymmetric multilayer composite billets is shown in Figure 4.37. Considering now the extrusion of these billets through wedge rigid dies, a theoretical estimation of the overall strength of the cladded/extruded rods, based on the concept of the formed "multilayer" bodies, suggested by some of the authors (see Reference [16]), is modified to properly accommodate the present conditions and it is outlined below.

The metal-sheathed ceramic rod is considered as an axisymmetric two-layer body with the outer metal and the inner ceramic material layers denoted by *a* and *b*, respectively. Note that the intermediate layer, representing the narrow transition zone at the interface, is considered very small and, therefore, it may be ignored; see the remarks made above and also Reference [15]. The situation existing in the die during extrusion in every pass is shown in Figure 5.5. The yield criterion for axisymmetric situations and for each material may be written in the form:

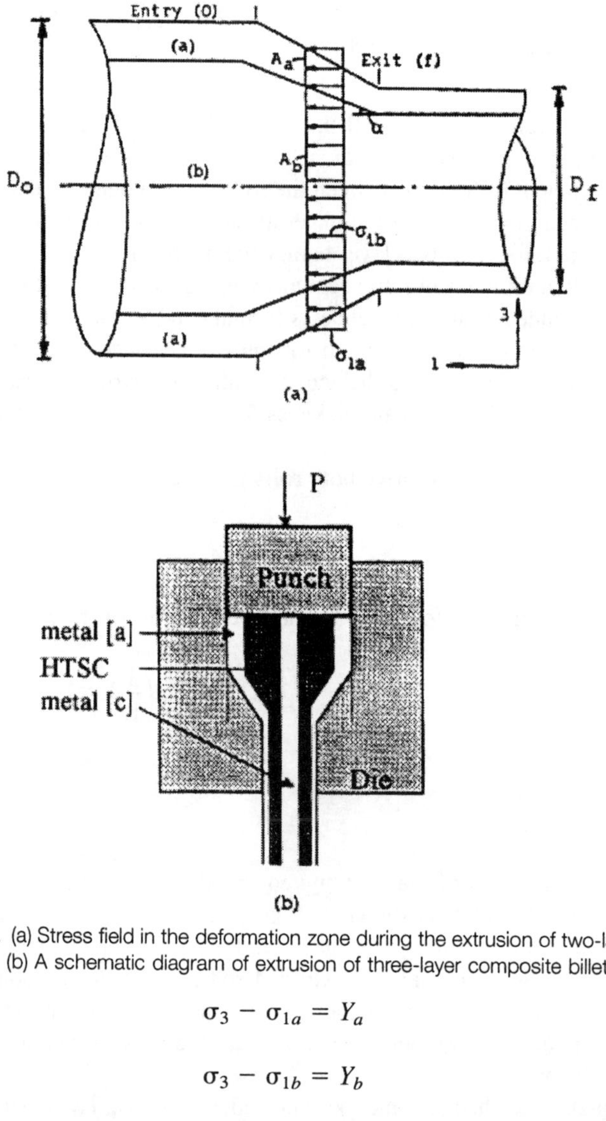

FIGURE 5.5. (a) Stress field in the deformation zone during the extrusion of two-layer composite billets. (b) A schematic diagram of extrusion of three-layer composite billets.

$$\begin{aligned} \sigma_3 - \sigma_{1a} &= Y_a \\ \sigma_3 - \sigma_{1b} &= Y_b \end{aligned} \tag{5.10}$$

Assuming that the horizontal stresses σ_{1i} $(i = a, b)$ are distributed uniformly over the corresponding vertical sections and each of them is the sum of the stresses developed because of conventional extrusion, x and of the transmitted shear stress, y_i at the interface between layers; it yields

$$\begin{aligned} \sigma_{1a} &= x - y_a \\ \sigma_{1b} &= x - y_b \end{aligned} \tag{5.11}$$

The effective yield stress, Y^* for the bimetallic extrusion may be taken, when the Tresca criterion is considered as:

$$Y^* = \sigma_3 - x \tag{5.12}$$

and considering also that the stresses y_a and y_b must be in equilibrium, i.e.,

$$y_a A_a = y_b A_b \tag{5.13}$$

by combining Equations (5.11)–(5.13), the effective yield stress of the bimaterial can be obtained as

$$Y^* = \frac{Y_a A_a + Y_b A_b}{A_a + A_b} \tag{5.14}$$

It may be emphasized that if Y^*_o and Y^*_f are the effective yield stress at the entry and exit, respectively, the mean effective flow stress of the bimaterial, $\overline{Y}$, when it has undergone a total strain $\epsilon = \ln R = \ln[1/(1 - r_T)]$, where R is the extrusion ratio and $r_T = (A_o - A_f)/A_o$ is the total reduction in the cross-sectional area of the rod, can be assumed as

$$\overline{Y} = \frac{Y^*_o + Y^*_f}{2} \tag{5.15}$$

By applying the classical "friction-hill" method, properly modified for extruded bimaterials, the punch load can be obtained [16]

$$P = \overline{Y} A_o \left[\left(1 + \frac{\tan \varphi}{\mu} \right) (R\mu \cot \varphi - 1) + \frac{2L}{D_o} \right] \tag{5.16}$$

where $\overline{Y}$ is the mean effective yield stress of the bimaterial, L the billet length in the container, D_o the initial diameter, A_o the cross section of the billet at the entry, φ the semiapical angle of the conical die and μ the coefficient of friction at the outer layer of the billet/die contact area.

In Figure 5.5(a) the extrusion through conical dies of a three-layer composite billet is considered. The metal-sheathed ceramic billet with a small diameter mandrel placed along its axis of symmetry is formed by cladding and compaction according to the configuration shown in Figure 4.37.

Following the analysis for the extrusion of bimaterials outlined above, taking also into account the assumptions made, from Equation (5.14) an effective yield stress, Y^* of the three-layer material may be suggested as

$$Y^* = \frac{Y_a A_a + Y_b A_b + Y_c A_c}{A_a + A_b + A_c} \tag{5.17}$$

and from Equation (5.15) the related mean effective yield stress, $\bar{Y}$. Subsequently, the punch load for the present case can be expressed from Equation (5.16), where $\bar{Y}$ denotes the mean effective flow stress of the three-layer material.

UPSETTING

Upsetting of cylindrical billets or plane-strain rectangular strips of the multilayer composites is used for the evaluation of their overall strength by determining their stress-strain characteristics. Tests are performed on standard universal testing machines at temperatures and strain rate conditions relevant to the actual forming ones. The mean effective yield stress of the multilayer materials, $\bar{Y}$, can be obtained by measuring the area under the stress-strain curve and dividing it by the corresponding amount of strain imposed. Typical stress-strain curves of metal-sheathed ceramic HTS, e.g., Ag/YBCO composites, are shown in Figure 5.6 [10]; see also below.

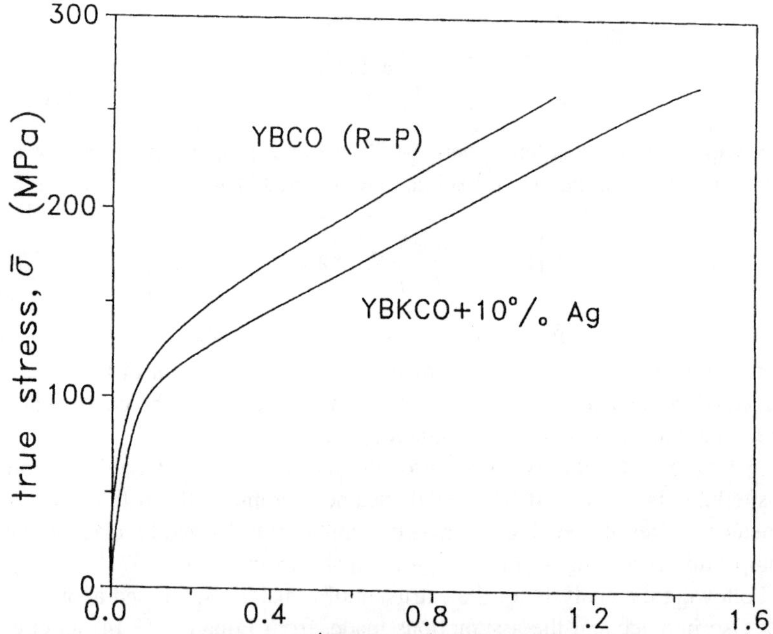

FIGURE 5.6. Stress-strain compression curves of silver-sheathed/YBCO billets at 450°C temperature. YBCO (R-P) is the commercial Rhone-Poulenc powder.

5.4 MANUFACTURING OF STRIPS AND TAPES

The main processing techniques encountered to the fabrication of strips and tapes (foils) are:

- the powder-in-tube (PIT) process, related to the forming of noncompacted ceramic core plates;
- the rolling of explosively cladded/compacted multilayer composites, related to the forming of precompacted ceramic core plates.

5.4.1 PIT Process

The powder-in-tube method, as mentioned above in Section 5.2, is widely used for industrial-scale production of superconducting tapes, see References [3], [4] and [17]–[23]. The following processing stages may be listed:

- Packing of the superconducting powder into a metal tube, followed by vibration, necessary for the homogenization of the powder. Subsequently, a punch compresses statically the powder to reduce the interparticle voids and, finally, a seal is placed at the opened end of the tube.
- Drawing of the metal-clad superconducting tube to form a thin rod.
- Rolling (1st pass) of the rod to form a rectangular tape.
- Intermediate heat treatment (solid-state or partial-melt) for sintering the ceramic core.
- Rolling (2nd pass) to produce the final thickness.
- Final heat treatment to re-innovate the superconducting properties and establishing high-bulk density.

These processing steps may be repeated many times in an industrial production line to obtain long Ag-clad superconducting tapes. Note that rolling can be performed in cold [21] or in hot conditions [22, 23]. A typical process phase diagram for the production of silver-sheathed 2212-BSCCO tapes is shown in Figure 5.7, reproduced from Reference [20]. Texture development and inner core discontinuities, e.g., the sausaging effect, indicating a highly oriented core structure of this material are shown in Figure 5.2(a) and (b), respectively.

5.4.2 Rolling of Explosively Cladded/Compacted Metal-Sheathed/HTS Multilayer Plates

This manufacturing sequence is used for fabricating metal-sheathed superconducting components, see References [5] and [6].

After the explosive compaction/cladding of the metal/YBCO and metal/BSCCO components (see Figures 4.22 and 4.23 of Chapter 4), of dimensions

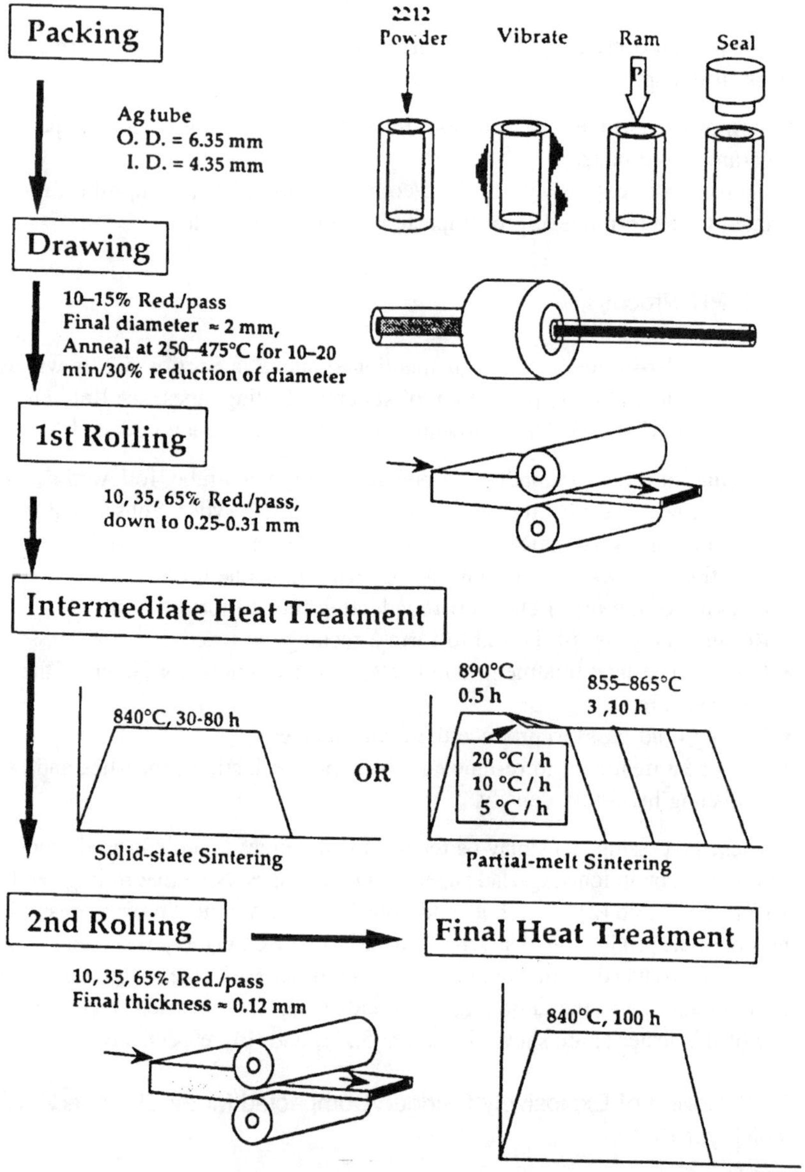

FIGURE 5.7. A flow chart of the PIT process (from Reference [20]).

200 × 40 mm and a certain thickness were cold rolled in successive passes to a final thickness reduction to produce thin superconducting strips. Rolling of the rectangular plates was performed on an experimental two-high rolling mill, properly instrumented for roll force and torque measurements, at a constant speed of 4 m/min between two steel rolls of 200-mm diameter and 100-mm barrel length. All rolling passes were performed in "dry" condition, whereas strips were subjected to a stress-relieving annealing at 550°C for 5 min, after the intermediate passes.

METAL/YBCO STRIPS

The initial thickness of the silver-clad/YBCO plates was 7 mm and after successive passes of total thickness reductions $r_T = 50\%$, 61% and 73% rolled strips of 2-mm thickness were fabricated. Stress relieving was performed at 500°C, after the 61% reduction was attained. The powder material in this case was the initial explosively compacted 2nd type powder of YBKCO system mixed with 10% Ag and the metallic-clad flyer and parent plates were made of silver; see Section 4.5.1(a).

Roll force and torque diagrams, recorded during rolling after a total reduction r_T equal to 61% and 73% are shown in Figure 5.8(a) and (b), respectively.

As mentioned above, failures occurring during compression working processes, such as rolling, are predominantly caused by secondary tensile stresses. Moreover, problems and difficulties arising from previous process, i.e., the explosive cladding, associated with certain structural changes and material properties of the composite strip, are sometimes intensified during the subsequent rolling process. However, defects, such as delaminations of ceramic from silver plates and alligatoring, occurred during rolling, were not observed, but macro- and microcracking were developed after the final rolling pass.

Longitudinal sections of the strips after 50% and final 73% total reductions were examined by scanning electron microscopy, see Figure 5.9. In both cases, a well-bonded silver/ceramic interface was revealed. Note that, in strips, after a reduction of 73%, a longitudinal crack was developed, propagating almost parallel to the rolling direction, see Figure 5.9(c) and (d). This crack constitutes an intergranular failure and may be attributed to the brittle response of the ceramic core because of the presence of tensile residual stress fields. Defects, but less profound, were also observed after the intermediate rolling; compare Figures 5.9(a) and (b) to 5.9(c) and (d), respectively.

The microstructure of the ceramic core of the above-mentioned samples is shown in Figure 5.10. It consists of irregular shaped ceramic grains without any preferred orientation, formed from the initially compacted grains after intense fragmentation and some frictional movement during the cold rolling. The reduction of grain size leads to the creation of new boundaries and to sub-

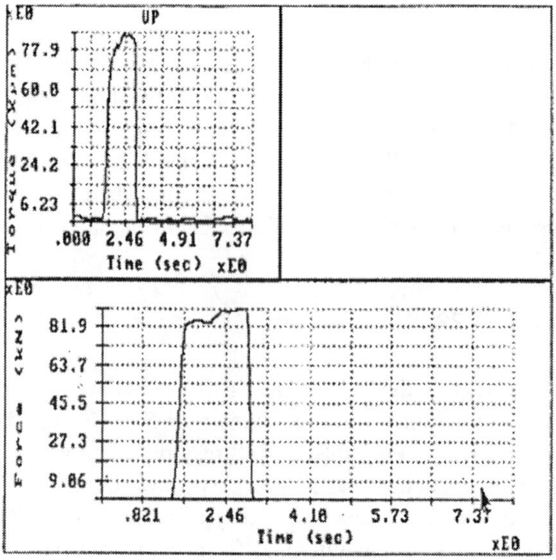

(a)

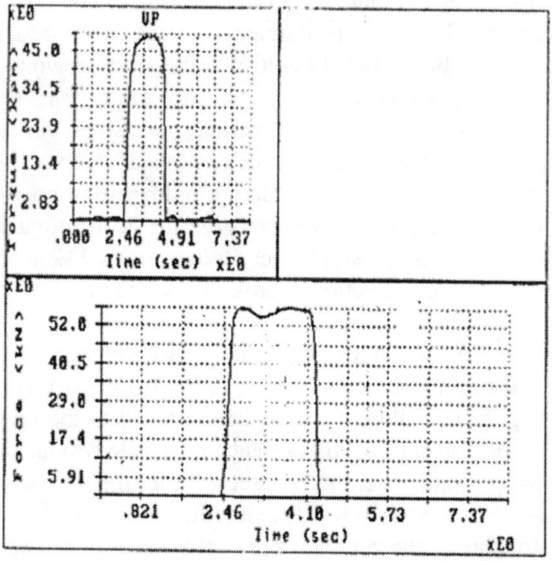

(b)

FIGURE 5.8. Distribution of roll force and torque during cold rolling of Ag/YBCO-cladded strips after a total reduction, r_T: (a) 61%; (b) 73%.

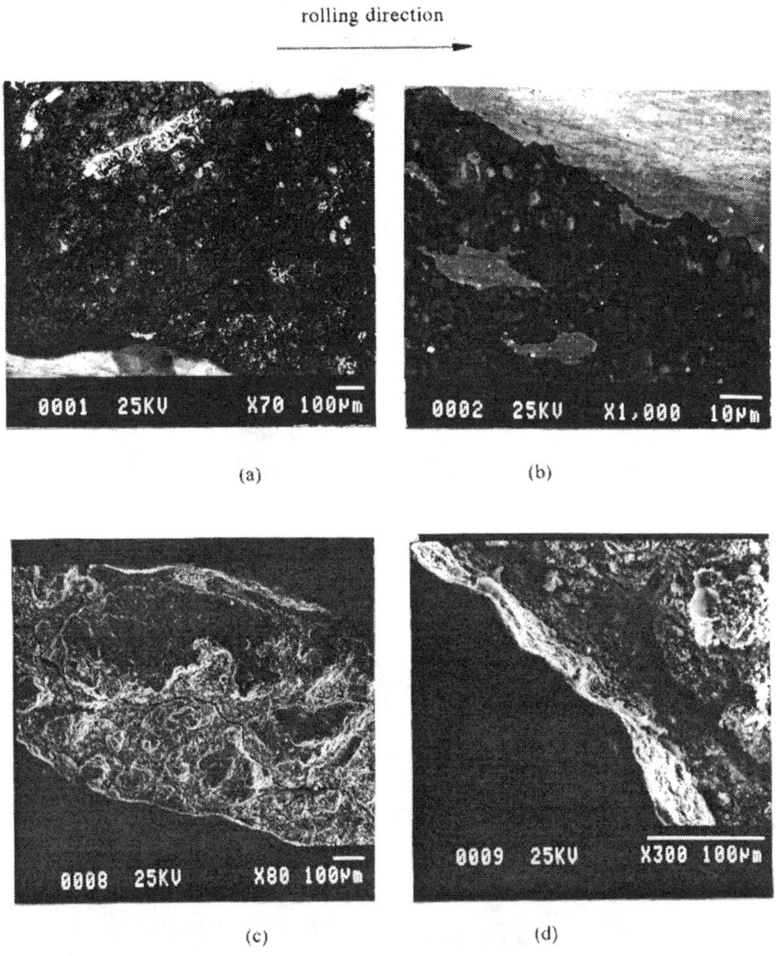

FIGURE 5.9. Scanning electron micrographs showing the bonding of the metal/ceramic interface and the cracking formation along longitudinal sections of the Ag/YBCO-rolled strips: (a) After r_T = 50%; (b) detail of the interface region of (a); (c) after r_T = 73%; (d) detail of the interface region of (c).

sequent partial elimination of the porosity. The new grain boundaries, associated with localized lattice disorder and secondary phase concentration, contribute to the enhancement of weak-link phenomena.

The superconducting transition temperature after the final rolling pass was slightly reduced to 85 K, whereas the width and height of the transition were more affected, see Figure 5.11; compare also with the magnetization curves of the cladded plates presented in Figure 4.36. The corresponding magneti-

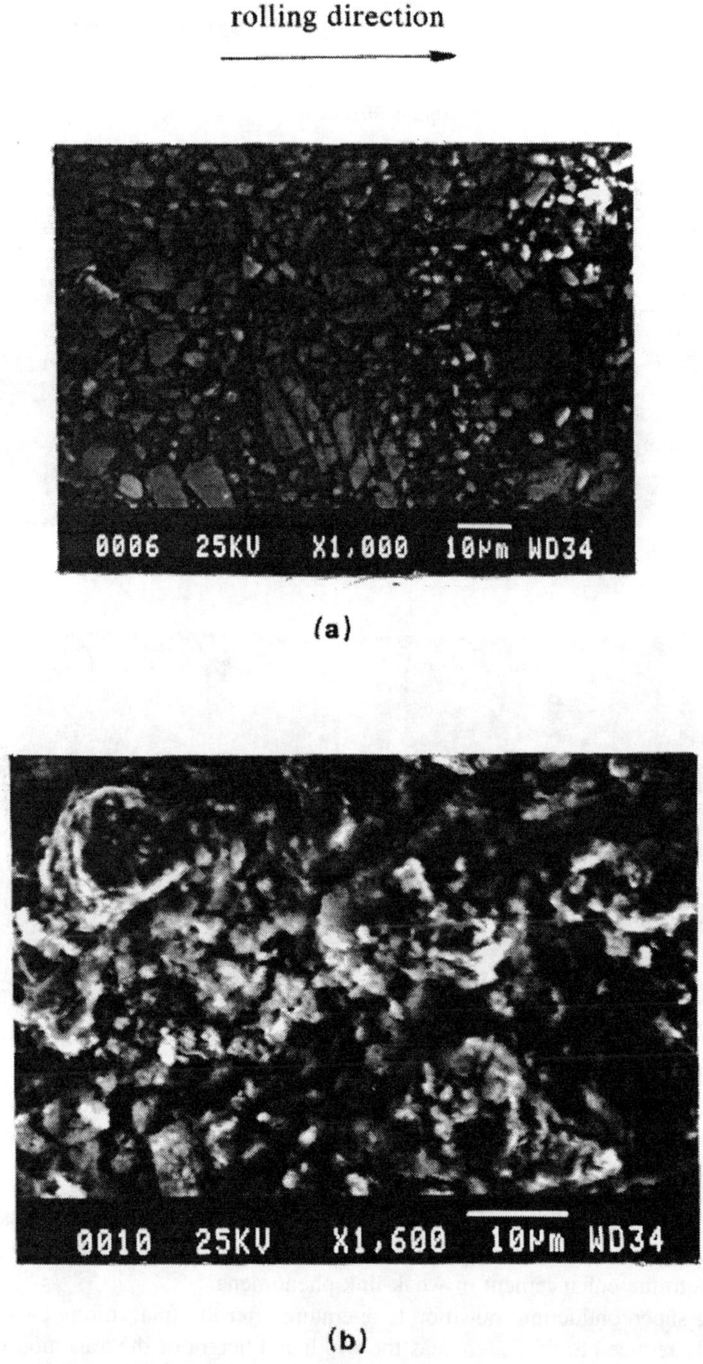

FIGURE 5.10. Scanning electron micrographs showing the microstructure of the internal YBCO ceramic core after rolling with (a) r_T = 50%; (b) r_T = 73%.

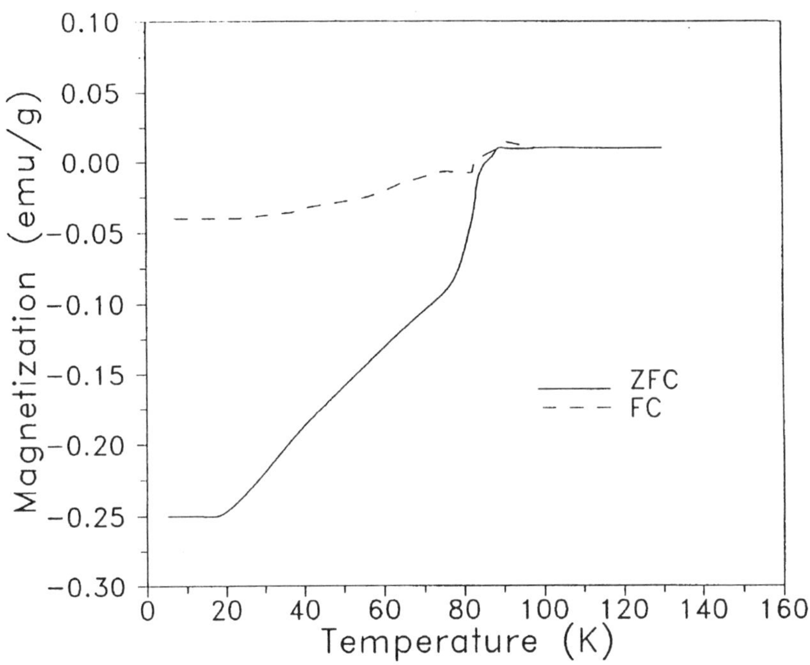

FIGURE 5.11. ZFC- and FC-magnetization curves for the rolled Ag/YBCO strip after a total reduction, r_T = 73%.

zation curves (ZFC and FC) became broader and smaller, as far as the Meissner signal intensity is concerned, indicating, therefore, further deterioration of the residual superconductivity, mainly attributed to strain-induced imperfections and to lattice distortion imposed by rolling.

METAL/BSCCO STRIPS

Multiple pass (12-pass) cold rolling was selected as the post cladding forming operation for further shaping the above-mentioned Ag/BSCCO-clad plates of the configuration shown in Figure 4.55, from their initial thickness of 4 mm to a final thickness reduction r_T = 78%, to produce superconducting strips of thickness about 1 mm. The strips were annealed at 550°C for 5 min, after two intermediate passes with r_T = 21% and r_T = 58%, respectively. Postfabrication heat treatment at 850°C for 24 h in air was performed after an intermediate rolling pass, r_T = 50%, and after the final rolling pass, r_T = 78%, followed by quenching in the furnace, to renew the superconducting properties of the produced components. Details about the materials used and the fabrication stages are tabulated in Table 5.1.

TABLE 5.1. Phase Content and Superconductive Characteristics of the Ag/BSCCO Materials at the Various Stages of the Fabrication.

	Critical Temperature or Temp. Width		Phase Content				
Fabrication Stage	T_c (K) T_{c1}/T_{c2}	ΔT_c (K) $\Delta T_{c1}/\Delta T_c$	(2223) % w/w	(2212) % w/w	(2201) % w/w	Ca_2PbO_4 % w/w	Ca_2CuO_3 % w/w
Initial powder	108/86	12/8	78.0	12	9.0	<1	—
Explosively compacted powder	110/87	20/10	67.0	21	11.0	<1	—
Rolled strip (r_T = 50%)	75	40					
Rolled strip (r_T = 78%)	70	50					
Heat-treated strip (r_T = 50%)	115/85	20/10	24.5	63	0.5	—	12.0
Heat-treated strip (r_T = 78%)	110/84	20/8	30.0	61	0.4	—	8.6

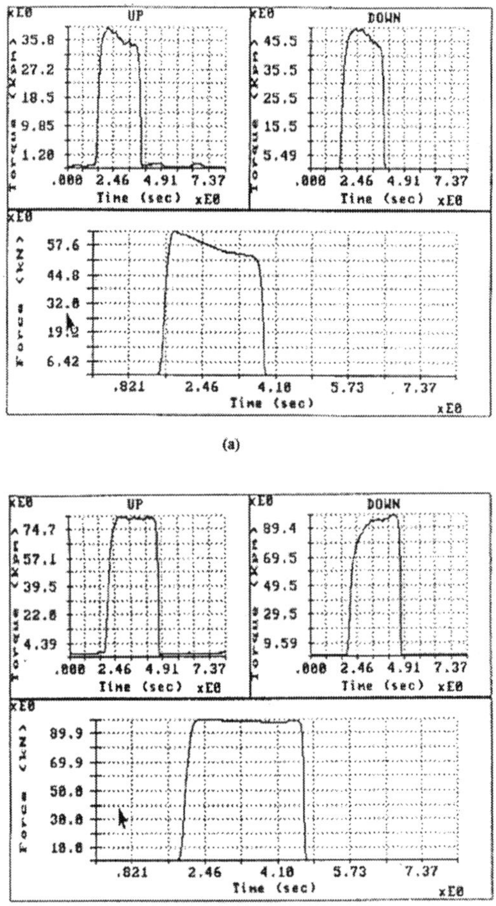

FIGURE 5.12. Distribution of roll force and torque during cold rolling of Ag/BSCCO-cladded strips after a total reduction, r_T: (a) 74%; (b) 78%.

The roll force and torque distribution for the last two rolling passes, r_T = 74% and 78%, are shown in Figure 5.12. Typical lateral spread, due to the inhomogeneous plastic deformation of the rolled strip, is indicated in the macrograph shown in Figure 5.13(a). However, characteristic rolling failures, such as wrinkling, aligatoring, edge cracking, sausaging etc., were not observed. Note also that problems and difficulties arising from the previous process, i.e., the explosive cladding, associated with microscopic interface fractures (see Figure 4.57) can be healed by the application of compressive stress fields resulting in the re-establishement of the Ag/ceramic bonding integrity after the

FIGURE 5.13. (a). Macrographs showing lateral spread of the Ag/BSCCO-rolled strip; (b) a longitudinal section showing the bonding integrity after a total reduction, r_T = 78%.

final rolling pass, see Figure 5.13(b). The thickness of the ceramic layer varied between 60 and 120 μm. Microhardness measurements, performed on the rolled strips, indicated a softening of the material, compared with the clad plates, due to the applied stress relief annealings, see Figure 5.14.

Scanning electron micrographs of the rolled strips after an intermediate rolling pass, $r_T = 50\%$, and after the final rolling pass, $r_T = 78\%$, reveal a gradual grain alignment to the rolling direction, see Figures 5.15 and 5.16. This preferred orientation and grain elongation were intensified after the final rolling stage; compare Figure 5.15 with Figure 5.16. Note that grain orientation may cause an increase in critical current density, J_c; superconducting tapes or ribbons, produced by cold rolling, possess a very high critical current density, mainly because of texturing by mechanical alignment.

Measurements of ac-magnetic susceptibility in relation to temperature reveal that a substantial change of the superconducting properties took place after rolling, see Figure 5.17(a). The residual superconductivity was measured after an intermediate rolling pass, $r_T = 50\%$, and after the final rolling pass, $r_T = 78\%$, see Figure 5.17(a). Broad transitions were observed at 75 K with $\Delta T_c = 40$ K and at 70 K with $\Delta T_c = 50$ K, respectively. Although no con-

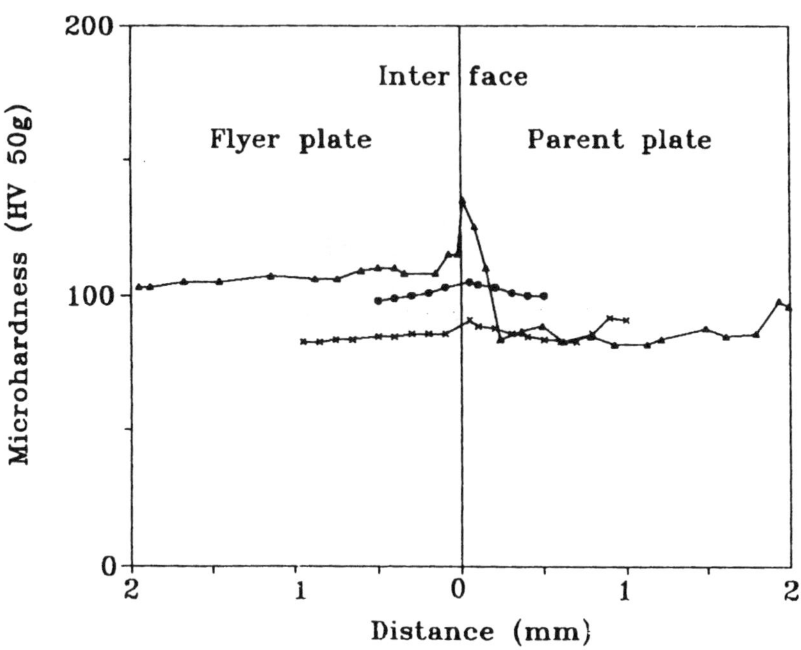

FIGURE 5.14. Microhardness of the silver as a function of the distance from the interface; –Δ–, as cladded; –×–, cladded and rolled ($r_T = 50\%$); –•–, cladded and rolled ($r_T = 78\%$).

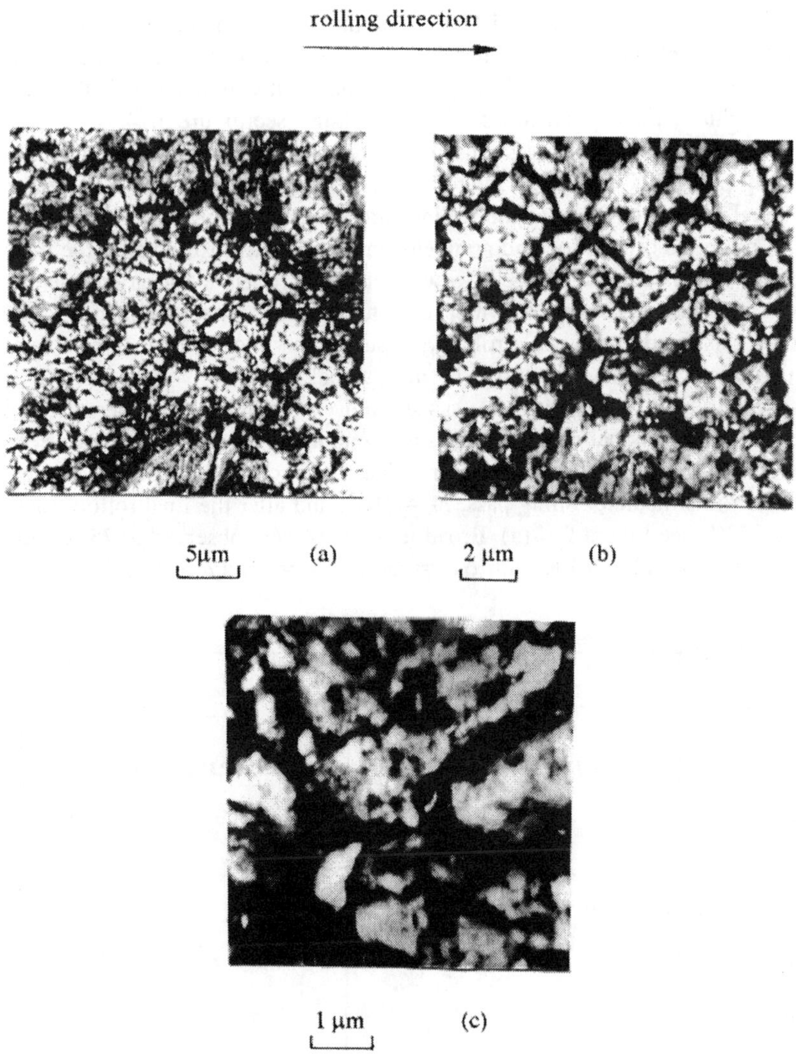

FIGURE 5.15. Scanning electron micrographs showing the microstructure of the BSCCO ceramic core after a 50% reduction.

siderable phase transformation is expected, because of the relatively low temperature, this deterioration of the superconductivity may be attributed, mainly to subsequent lattice defects that resulted from the excessive plastic deformation imposed during rolling.

Post fabrication heat treatment between 840 and 850°C is necessary to fill the oxygen vacancies and to make the lattice more perfect, by causing diffusive processes at the atomic scale, resulting, therefore, in the regaining of the

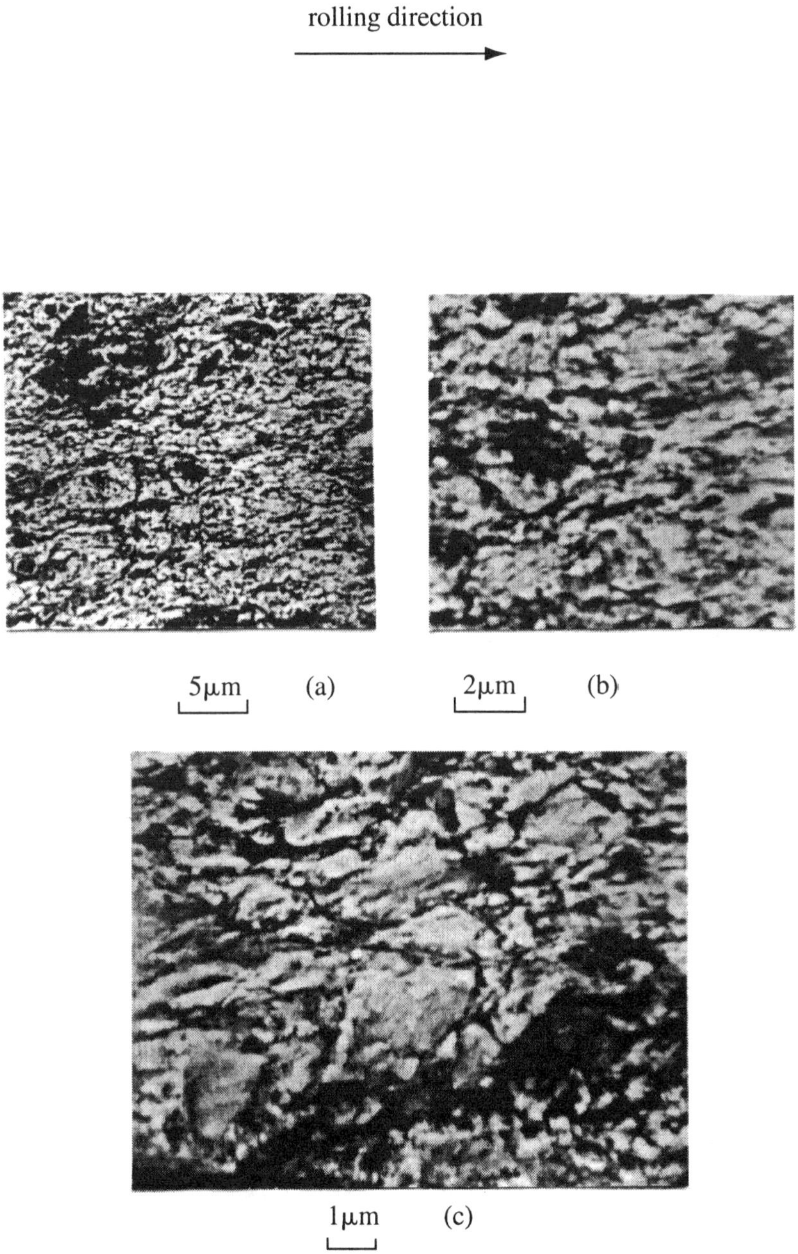

FIGURE 5.16. Scanning electron micrographs showing the microstructure of the BSCCO ceramic core after a 78% reduction.

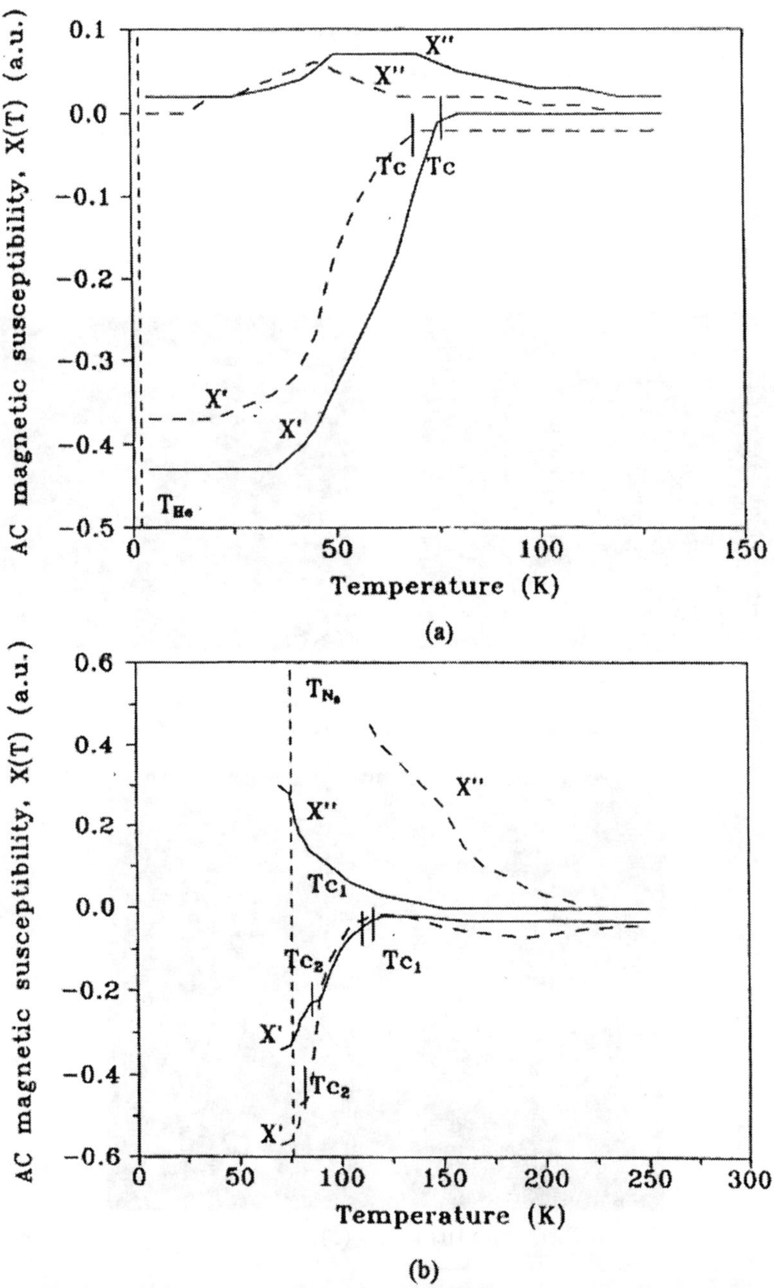

FIGURE 5.17. The ac-susceptibility curves of the (a) rolled and (b) heat-treated strips for reductions; solid line, r_T = 50%; dashed line, r_T = 78%.

superconducting properties. The formation kinetics of the 2223 phase is improved by heat treatment, leading to the enhancement of the superconductivity of the fabricated strips.

XRD patterns of the heat-treated strips, after an intermediate rolling pass, $r_T = 50\%$, and, after the final rolling pass, $r_T = 78\%$, indicated the existence of the three phases of the Bi-Pb-Sr-Ca-Cu-O system, namely the 2223, the 2212 and the 2201 phases, as well as the Ca_2CuO_3-phase; see Figure 5.18(a) and (b). Quantitative analysis, performed by applying the Rietveld method, predicted 63% w/w for the 2212 phase, 24.5% w/w for the 2223 phase, 12% w/w for the Ca_2CuO_3 phase and 0.5% for the 2201 phase in the former case, whereas in the latter case the related quantities obtained were 61% w/w for the 2212 phase, 30% for the 2223 phase, 8.6% for the Ca_2CuO_3-phase and 0.4% for the 2201 phase. Intense transformation of the 2223 phase, 42.5% and 37%, respectively, and of the 2201 phase, 10.5% for both cases, to the low-T_c 2212 phase and to Ca_2CuO_3 compound took place during the heating of the strips at 850°C for 24 h; see Figure 5.18(a) and (b). The Ca_2CuO_3 phase is a side product of the partial formation reaction of the 2223 phase.

Scanning electron micrographic observations of the heat-treated ceramic core ($r_T = 78\%$) revealed smooth glass-like areas formed because of the partial melting and subsequent resolidification processes during the thermal cycling, see Figure 5.19(a) and (b). The microstructure of the heat-treated strip, after the final rolling pass, mainly consists of darker flake-like crystallites incorporated in the lighter matrix, see Figures 5.19(c) and (d). Taking into account the XRD patterns of this strip, outlined above, and also similar remarks reported in References [24] and [25], it may be concluded that flake-like grains constitute the high-T_c phase grown from the liquid phase. The Ca_2PbO_4 phase, existing after the explosive compaction, melts at 820°C, helping, therefore, in the formation of the liquid phase during firing and in the crystallization of the high-T_c phase.

The ac-magnetic susceptibility measurements indicated the improvement of the superconducting properties after heat treatment, see Figure 5.17(b). In the case of the heat-treated strip after the intermediate rolling pass, two transition regions were observed, namely at $T_{c1,onset} = 115$ K, with ΔT_{c1} approximately 20 K, which resulted from the 2223 phase, and at $T_{c2,onset} = 85$ K, with ΔT_{c2} approximately 10 K, which resulted from the 2212 phase, see Figure 5.17(b) and Table 5.1. In the second case of the heat-treated strip, after the final rolling pass, two transition regions were also observed; a relatively sharp transition at $T_{c1,onset} = 110$ K, with ΔT_{c1} approximately 20 K, resulted from the 2223 phase and at $T_{c2} = 84$ K, with ΔT_{c2} approximately 8 K, resulted from the 2212 phase, see Figure 5.17. Note that although the 2212 phase exists in both heat-treated strips, in a greater percentage than the 2223 phase, the superconducting volume fraction of the high-T_c phase is greater than that of the low-

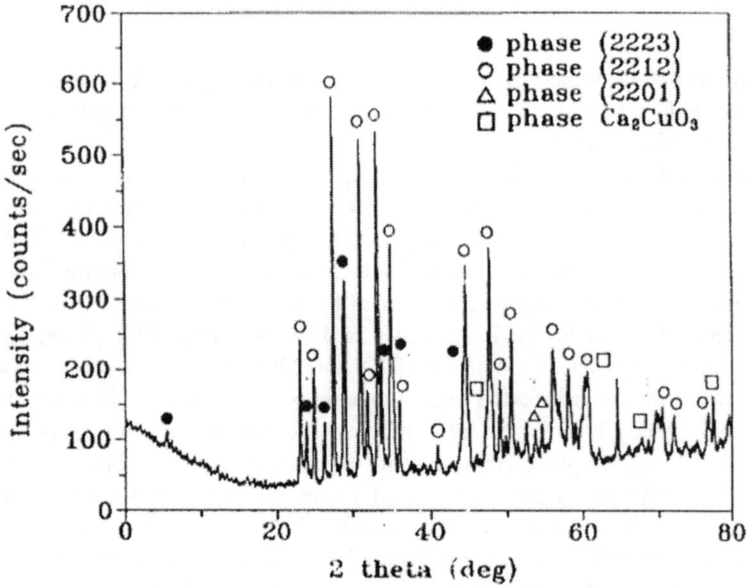

(a)

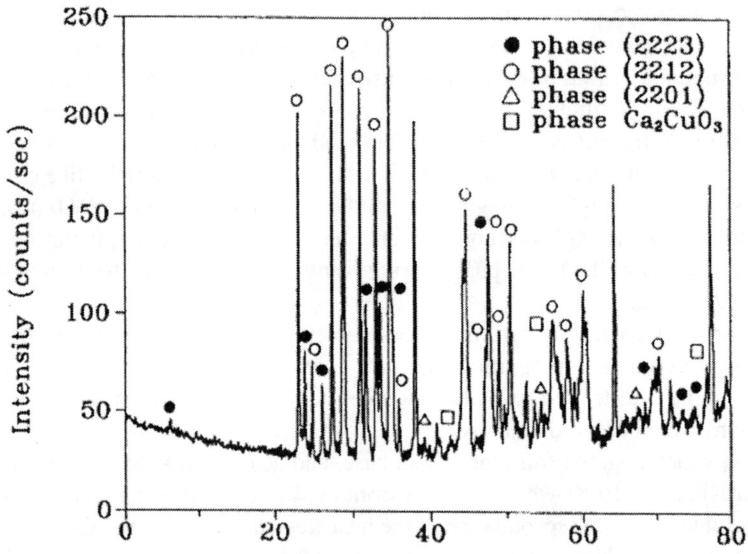

(b)

FIGURE 5.18. XRD patterns of the heat-treated Ag/BSCCO strips after a total thickness reduction: (a) 50%; (b) 78%.

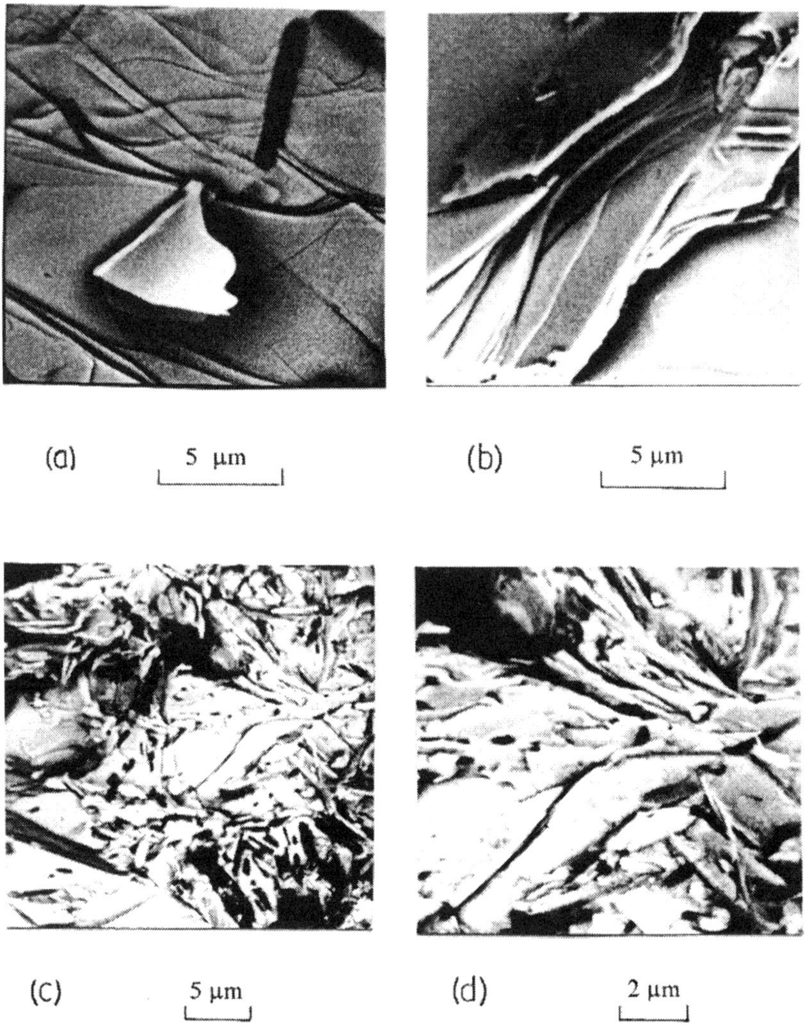

FIGURE 5.19. Scanning electron micrographs of the heat-treated Ag/BSCCO ceramic after a total reduction, r_T = 78%, showing (a) and (b) "glassy"-like microstructure; (c) flake-like 2223 phase crystals; (d) detail of (c).

T_c phase, as obtained from the diamagnetic shielding signals ratio, see Figure 5.17(b).

Preliminary measurements of dc-resistivity in relation to temperature for both strips exhibited inflection points near a temperature of 110 K, see Figure 5.20. These inflection points may indicate the onset critical temperatures for each case.

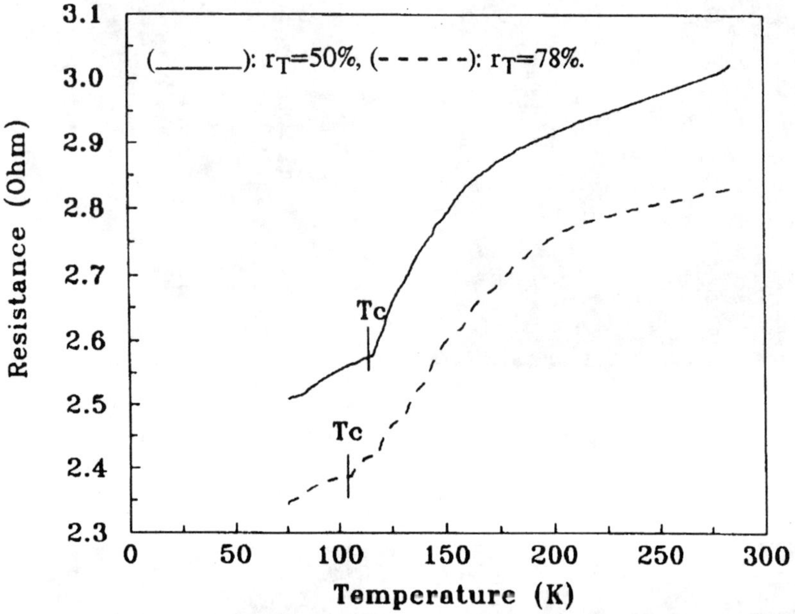

FIGURE 5.20. Resistivity curves of the heat-treated Ag-sheathed/BSCCO ceramic strips after a total thickness reduction; ——, r_T = 50%, ————, r_T = 78%.

5.5 MANUFACTURING OF RODS AND WIRES

Several fabrication techniques are used for the processing of high-T_c ceramic superconductors into rods and wires. Significant ductility is required for shaping, which is not possessed by the ceramic oxides alone. The main configurations of fabricated axisymmetric high-T_c superconducting components, in the form of thin rods and wires, may be classified as:

- Nonsheathed HTS axisymmetric component, involving:
  —extrusion of ceramic oxides with binders, plasticizers and wetting agents and extrusion of compacted metal-oxide microcomposites;
  —melt texturing and solidification techniques;
  —reaction texturing of metal-oxide microcomposites.
- Coated metal fibers involving deposition forming.
- Metal-sheathed HTS rods and wires involving:

  —extrusion of powder-in-tube metal-oxide powders;
  —extrusion and profile rolling of explosively compacted metal/superconducting billets.

5.5.1 Non-Sheathed HTS Wires

EXTRUSION OF CERAMIC OXIDES

The powder/binder or plastic extrusion process is probably the main production technique for unsheathed superconducting axisymmetric components. It involves preparation of a paste by mixing the ceramic powder with a certain percentage of an organic binder, e.g. ethyl cellulose. Powders of the YBCO system and the BSCCO compound are used, see References [26] and [27]. The powder/binder mixture is sufficiently ductile, and, therefore, it can be plastically shaped to geometries useful for wire or current lead applications by using conventional wire-drawing or extrusion processes. The deformation processing is then followed by a firing process for the binder removal and by sintering for powder densification.

Warm and hot extrusion was used to produce flexible superconductive wire from metal-oxide microcomposites. The microcomposites of silver oxide/ YBCO powders, produced by powder mixing, were statically compacted in warm or hot conditions, and the billets, after sintering, were extruded at elevated temperatures to produce thin wires. Thermal treatment in flowing oxygen resulted in regaining the superconductive properties of the components; see References [28] and [29].

MELT-TEXTURING PROCESS

The melt-texturing (MT) technique is widely used for the fabrication of bulk $YBa_2Cu_3O_7$ ceramics because of the enhanced transport and magnetic properties of this material. Because of the yttrium diffusion-controlled growth mechanism of 123, the maximum allowable growth rate depends on the size and distribution of Y_2BaCuO_5 (211) precipitates. In addition, a slow cooling rate and a slow growth rate should be used to maintain the planar growth front. Among the many melt-texturing methods and modifications, seeded directional solidification has the potential to produce long single-domain samples of large cross-sectional area with an intense orientation of the *ab* crystallographic planes along the sample axis. Increased progress has also been achieved in flux-pinning enhancement through the creation of artificial defects, e.g., the formation of 211 phase precipitates in 123 matrix during melt texturing.

Melt texturing of the 123 compound results in a microstructure that consisted of large 123 plates, stacked along the *c*-axis direction. All the melt-texturing methods involve the melting of the 123 above its peritectic temperature, which is approximately 1010°C in atmospheric pressure. When 123 is heated above this temperature, incongruent melting occurs, leading to the for-

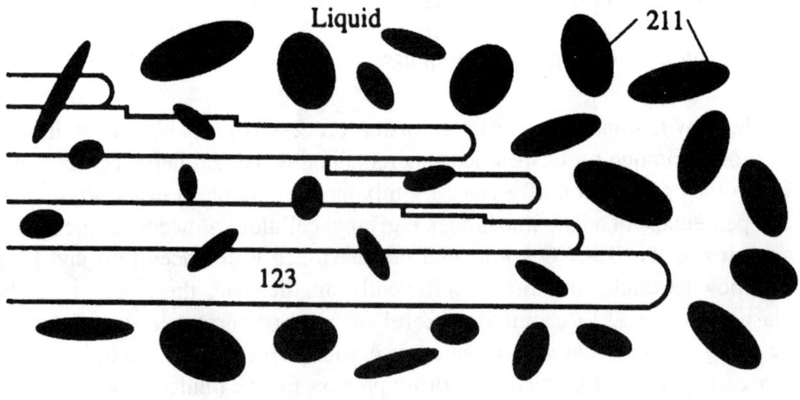

FIGURE 5.21. A schematic diagram showing the nucleation growth processes in melt-textured 123 (from Reference [30]).

mation of a solid (211) plus a liquid phase. In subsequent, carefully controlled, slow cooling of the semisolid melt, oriented-grained 123 morphologies are formed with 211 phase inclusions, distributed throughout the superconducting matrix. The nucleation growth mechanism is controlled by the Yttrium-diffusion front. A schematic illustration of nucleation growth processes during melt texturing is shown in Figure 5.21. To decrease the 211 particle size and to enhance the uniformity of their distribution, modified melt-texturing methods, e.g., the quench-melt-growth (QMG) and the melt-powder-melt-growth (MPMG), are used.

Large oriented-grained or even single-crystal samples can be fabricated by melt-texturing technique. Enhanced grain alignment and flux pinning obtained by melt texturing results in very high critical current densities, of the order of 10^5 A/cm^2 at 77 K. A comprehensive review of the various aspects of melt texturing process has been reported by Salama and Lee [30].

COMPOSITE REACTION TEXTURING PROCESS

Large bulk HTS components, mainly of the BSCCO system, can be fabricated by using the composite reaction-texturing technique (CRT). According to this technique a composite preform, seeded with an aligned distribution of MgO fibers, is prepared by elastomer processing or by tape casting, and reacted in three stages. The first stage is binder burnoff carried out in air; the optimum heating sequence depends on sample cross section and results in a complete removal of the binder without the creation of serious macroscopic defects. The second stage involves densification in a pure oxygen atmosphere to >95% bulk density, followed by partial melting and cooling. Finally, the good current links, the oxygen stoichiometry and the ordered crystal structure can be established by postannealing in oxygen low pressure. Composite HTS conductors for electrical machines or current leads may be manufactured by using the CRT method. The critical current density of the CRT products varies between 10^3 and 10^5 A/cm^2 at 77 K [31].

5.5.2 Deposition Forming of Coated Metal Fibers

According to this coating process, the superconducting ceramic is continuously deposited at a standard deposition rate on a thin metal fiber. The processing route consists of the following stages [32]:

- Electrophoretic deposition of HTS on metal fiber: A nickel alloy substrate is fed into the coating chamber containing slurry of the YBCO material.
- Sintering: Following electrophoretic deposition of the YBCO particles from the slurry, the coated metal fiber passes through a sintering furnace at a temperature 950–1030°C in oxygen atmosphere for powder densification.
- Electrophoretic deposition of Ag on HTS coated fiber: Subsequent to the sintering, the HTS-coated fiber enters a silver-coating chamber containing a slurry of silver particles.
- Sintering: Another sintering in oxygen atmosphere at a temperature 850–950°C follows the Ag deposition process.
- Annealing: Low-temperature oxygen annealing at a temperature 400–500°C is used as the final production step for the establishment of the superconducting properties.

5.5.3 Metal-Sheathed HTS Rods and Wires

Forward extrusion or drawing through an axisymmetric die can be used as a primary process for forming thin rods from metal-superconducting powder-in-tubes, see Section 5.3.1. Forward cold or warm extrusion and profile rolling are also used as secondary processes for fabricating thin rods and wires from explosively compacted metal-sheathed superconducting billets.

COLD EXTRUSION OF METAL/YBCO RODS

The manufacturing stages for the forward extrusion of Ag/$YBa_2Cu_3O_7$ and St/potassium doped YBCO/Ag "green" billets are described in detail in References [7] and [8], respectively. Single-pass cold extrusion was performed on a 1000 MN hydraulic press, fully automated and equipped with force (piezoelectric crystal) and displacement (LVDT) measuring devices [7, 8]. Com-

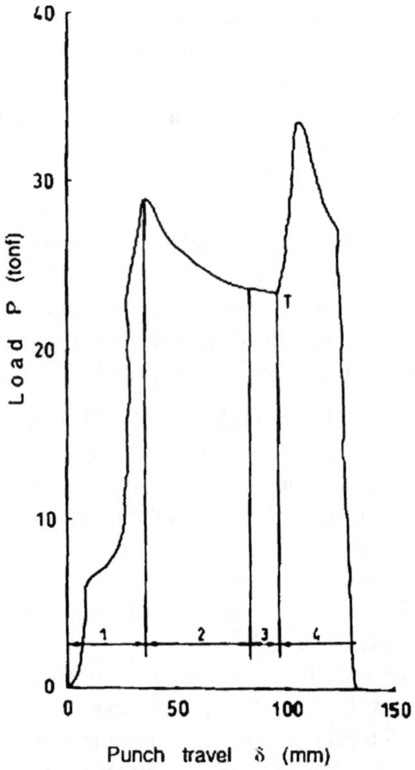

FIGURE 5.22. Load-punch travel diagram for Ag/YBCO billets compacted with cord-form explosive and extruded through a 90° steel die.

pacted Ag/YBCO billets of 18-mm diameter and St/YBKCO/Ag billets of 14-mm diameter were extruded at slow speed through 60°, 90° and 45° apex-angle steel dies, under lubricated conditions between the metal-sheathed billet, the steel extrusion container, the punch and the die by using a commercial lubricant, to fabricate rods of 10- and 8-mm diameter, respectively. A schematic diagram of the billet extrusion process is shown in Figure 5.5.

A typical extrusion load-punch travel diagram, as obtained from the autographic recorder, is presented in Figure 5.22. Longitudinal sections of the extruded specimens, along the horizontal axis of symmetry, examined by optical and scanning electron microscopy, revealed various types of macro- and microdefects, see Figures 5.23–5.25.

The punch load-punch travel, P-δ curve of the extruded composite rod with a ceramic core and silver sheath (see Fig. 5.22 [5.7]) follows, in general, the same pattern as the corresponding P-δ curves of metal billets forward extruded

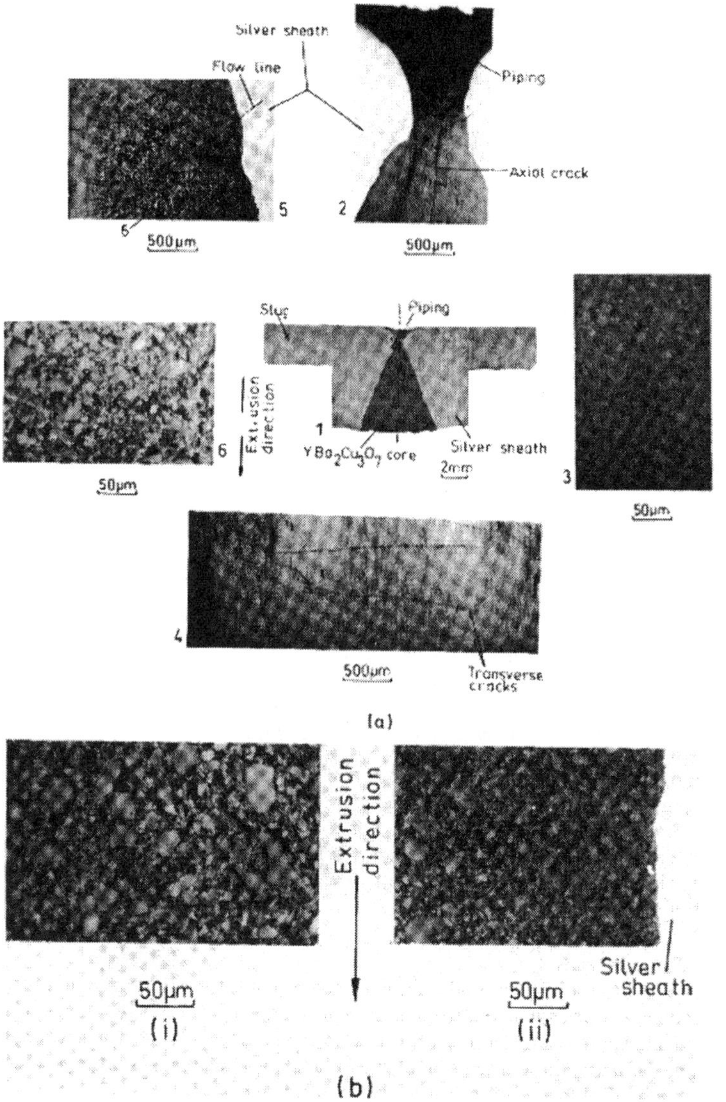

FIGURE 5.23. (a) Microstructural changes in a longitudinal section of a rod fabricated from a Ag/YBCO billet explosively compacted by using the cord-form explosive and extruded through a 90° die. 1, a schematic diagram showing the material morfology; 2, a photograph showing the piping defect and axial crack at the slug bottom adjacent to the punch. Micrograph showing 3, the high-density slug area; 4, transverse cracks observed at the high-density slug material; 5, the flow line formation in the slug; 6, grain fragmentation near to a flow line. (b) Micrographs showing the microstructure of the extruded rod after leaving the deformation zone near to (i) its axis of symmetry; (ii) the core/silver sheath interface.

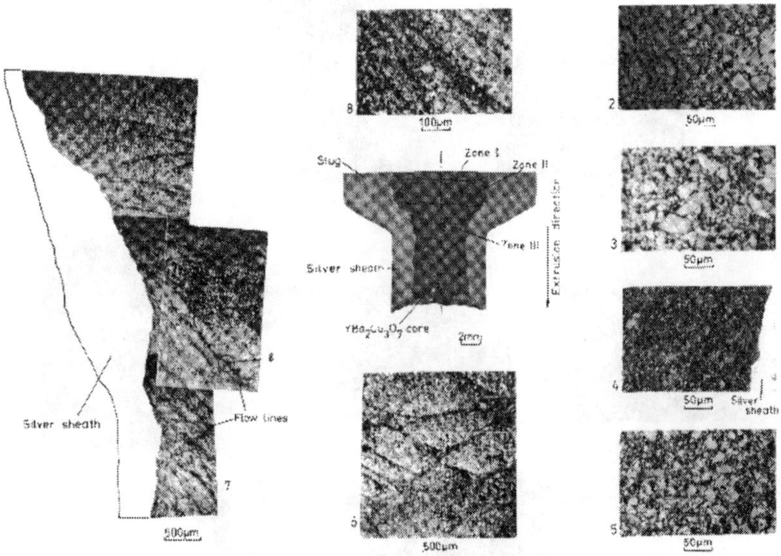

FIGURE 5.24. (a) Microstructural changes in a longitudinal section of a rod fabricated from a Ag/YBCO billet explosively compacted by using the cord-form explosive and extruded through a 60° wedge-shaped die. 1, a schematic diagram showing the damaged zones observed. Micrographs showing 2, the slug microstructure in Zone I; 3, the slug microstructure in Zone II; 4, the rod microstructure near the superconducting core/silver sheath interface; 5, the rod microstructure near the rod axis of symmetry; 6, the flow lines intersection on the rod axis of symmetry inside Zone III; 7, the direction of flow lines inside the wedge-shaped die near to the core/silver sheath interface; 8, the flow line microstructural characteristics of 7.

through dies of 90° die-angle [1]; the P-δ diagram may be made up of four regions; see Reference [7]:

- The transient or initial stage of extrusion (1), where all clearances between the extrusion chamber and the billet are taken up; when the maximum load has been reached extrusion proper begins.
- The steady-state phase (2), with the load falling steadily as the billet length reduces because of the continuous reduction in slug length remaining in the chamber.
- The unsteady stages (3) and (4), when the billet is short, leading to piping formation when the slug is sufficiently thin. The well-known piping defect in metal extrusion, i.e., the formation of a hole or pipe on the slug face adjacent to the punch on the center-line when the slug is sufficiently thin [1], was observed in the present case, see 2 in Figure 5.23(a); piping is usually associated with the bottom of the bowl-shaped portion of the load-punch travel curve, see Figure 5.22. The slug consists of highly dense material with an al-

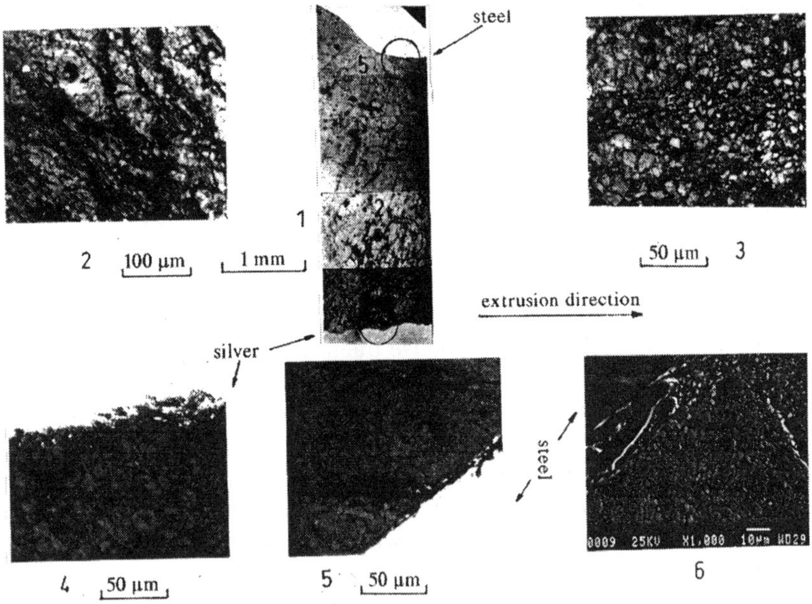

FIGURE 5.25. Optical and SEM micrographs showing the various defects formed in the region of the plastic zone of a St/YBKCO/Ag billet extruded through a 45° wedge-shaped die. 1, macrograph of the plastic zone; 2, network of shear cracks; 3, microcrystalline area; 4, detail of the Ag/YBKCO interface; 5, cracking occurred at the steel/YBCO interface; 6, metal debris detached from the steel wall due to the intense friction phenomena occurred during extrusion.

most complete lack of porosity, see 3 in Figure 5.23(a), but with the appearance of axial and transverse cracks, as indicated in 2 and 3 of Figure 5.23(a), respectively. Flow lines were formed also in the slug region initiating from the superconducting core/silver sheath interface and spreading toward the rod axis, see 5 in Figure 5.23(a). This flow line formation is responsible probably for the particle fragmentation in shear and the relative movement of fragments and their rearrangement that is observed, see 6 in Figure 5.23(a).

- Note that the extruded rod is characterized by a ring-shaped region of small grains adjacent to the superconducting core/silver sheath interface and expanding approximately 0.5–1 mm into the rod core, whereas the central area of the core consists of large grains, see Figure 5.23(b).

In Fig. 5.24 a longitudinal section of the rod and the remaining slug produced by forward extrusion through a wedge-shaped die of 60° die-angle, of a compacted Ag/YBCO billet. Three distinct zones can be defined, taking into account the microstructural changes observed and the mechanical forces on the billet during extrusion; see Reference [7]:

- Zone I, in the remaining slug of the billet, is affected by the stress field induced into the compact body from the punch/billet interface. The compact in this zone exhibits a rather lower porosity than that of the compacted billet before extrusion. Powder particle fragmentation without relative fragment movement took place, probably because of the applied high compressive stress field.
- Zone II, also in the remaining slug of the billet, is separated from Zone I by a transverse crack. The grain fragmentation observed in the billet during the shock compaction and reported above is apparent, see in Figure 5.24. Note that the porosity is not reduced significantly, and particle fragmentation is not followed by relative fragment movement.
- Zone III is related to the extruded material inside the die that has been subjected to the extrusion loading and to the frictional conditions prevailing along the die face. Increased grain fragmentation followed by the relative movement of fragments is observed in this zone. These microstructural changes are associated with the formation of flow lines, i.e., ductile shear fractures (see 7 in Fig. 5.24), which start from the silver sheath/compact interface at about 45° to it and then extend inside the compact, intersecting each other at approximately 90° on the rod axis of symmetry, see also 6 in Figure 5.24. Microstructural details of a flow line path are shown in Figure 5.24; this kind of fracturing may be associated with trajectories of maximum shear stress (slip lines) and takes place under severe distortion of the powder particles in the direction of plastic flow. Severe particle fragmentation took place close to the interface between the superconducting core and the silver sheath of the rod, see 4 in Figure 5.24, the width of this damaged area being approximately 1 mm. Incomplete particle fragmentation and relative fragment movement are the main characteristics of the core material close to the rod axis and outside the ring-shaped zone mentioned above, see 5 in Figure 5.24. The intersection of flow lines on the axis of symmetry of the rod in Zone III results in the formation of macrovoids observed in the rod axis.

In extruded steel/YBKCO/silver billets through 45° wedge-shaped dies, a network of 45° shear cracks, relevant to slip-line fields, were developed and intensified near the center of the rod, leading, therefore, to crack propagation along the maximum shear stress direction, see 1 and 2 in Figure 5.25. Intense grain fragmentation, resulting in severe reduction of the grain size in this vicinity, as shown in 3 of Figure 5.25. Examination of the silver/ceramic interface indicated a good bonding and the absence of serious damages, see 4 in Figure 5.25. However, large-scale cracking was observed at the steel/ceramic interface, because of higher plastic deformation imposed, see 5 in Figure 5.25. Steel particles were completely detached from the corresponding interface, see

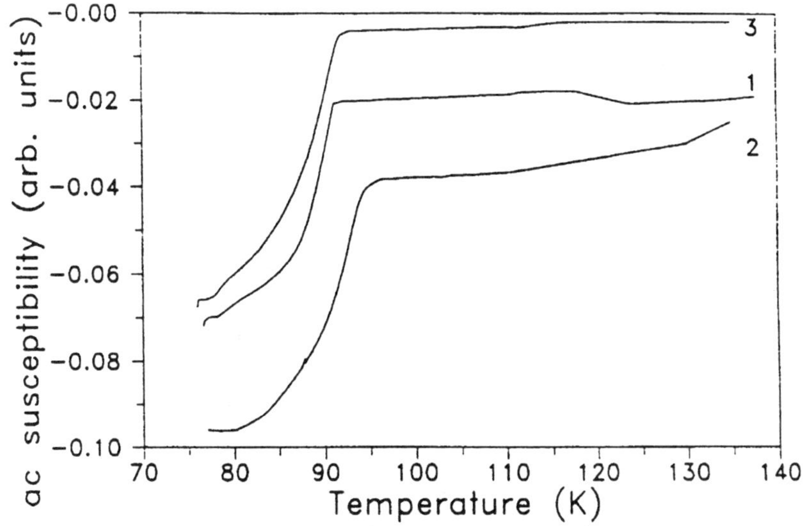

FIGURE 5.26. The ac-susceptibility curves (real part) for 1, the initial powder; 2, the explosively compacted billet; 3, the extruded St/YBKCO/Ag rod through a 45° wedge-shaped die.

6 in Figure 5.25, because of high friction that developed between the steel and the ceramic material during the plastic deformation.

Magnetic susceptibility measurements performed on the intitial powder, on compacted billets and in extruded St/YBKCO/Ag rods indicated that the normal/superconducting transition occurred at an onset temperature of 92 K, whereas the corresponding width was 14 K, see Figure 5.26 and Table 5.2. The relatively small deterioration of the residual superconductivity is attributed mainly to the lattice distortion and microstructural defects that resulted from intense cold deformation. Judging from the superconducting properties of the ceramic material at the various stages of the fabrication, it can be concluded that, in general, superconductivity is maintained after processing and subsequent postprocessing heat treatment in oxygen atmosphere may not be necessary.

WARM-EXTRUSION OF METAL/YBCO RODS

Superconducting components for high-power technological applications (rotating machines, generators, bearings), should possess, apart from the necessary electromagnetic properties, satisfactory mechanical properties and structural integrity, associated with a sufficient formability. The brittle behavior of the ceramic core makes difficult the postcompaction deformation processing.

TABLE 5.2. Superconducting Properties of the Steel/YBKCO/Silver Axisymmetric Components at the Various Stages of Fabrication.

Material	$T_{c,onset}$ (K)	$T_{c,offset}$ (K)	ΔT_c (K)
Initial powder	90	81	11
Compacted billet	93	80	13
Extruded (45° die-angle) rod	92	78	14

In addition, many macrostructural defects and failures, such as radial/longitudinal cracking and delamination, are developed during cold working of the compacted ceramics. These failures may be reduced, or even eliminated, by using warm working techniques in the vicinity, or even below, the recrystallization temperature (0.6–0.7T_m); creep flux flow and internal friction mechanisms, in relation to the plastic work imposed, improve the formability of the ceramic core, leading to crack closing, to the minimization of porosity and to the subsequent improvement of mechanical properties.

For the deformation processes at high temperatures, creep flow may result in high-density materials and, in many cases, e.g., BSCCO-layered compounds, in ideal texturing by flattening the initial grains. Besides the preferred crystallographic orientation, appropriate plastic deformation accompanied by heating at temperatures high enough, say above 900°C, may lead to the increase of the defect density of the ab-planes, creating mainly dislocations and stacking faults and to the subsequent enhancement of the J_c [33].

Warm working combines the advantages of both hot and cold working in one process. Improved dimensional accuracy, decrease of surface roughness, i.e., enhanced surface integrity, and lower energy consumption constitute the main advantages of warm working, compared with hot working, whereas increased ductility is its main advantage over cold working. Warm working of compacted HTS billets, e.g., by extrusion at 600°C, may lead to possible dynamic recrystallization phenomena, resulting in the decrease of the initial average grain size, by mechanisms of nucleation and grain growth. However, the absence of oxygen atmosphere, during warm working, causes severe deterioration of superconductivity, due to oxygen loss and the subsequent appearance of the tetragonal phase and to the formation of normal secondary phases. In the present case, by optimizing the extrusion temperature at 470°C, and balancing, therefore, the oxygen diffusion in and out of the sample and suppressing the secondary phase formation, this degradation has been minimized. For this reason, the microstructure (phase content) and the superconducting properties must be carefully monitored during fabrication.

Multiple-pass warm extrusion was performed on the same 1000 MN hydraulic press used for the cold-extrusion outlined in (a) above. Explosively compacted silver-sheathed billets of the YBKCO compound, with a 12-mm outer diameter and 7 mm inner (ceramic core) diameter, were extruded at 470°C through a 30° wedge-shaped split die (fabricated from hardened/tempered and nitrided tool steel), under lubricated conditions between the billet and the tooling, by using a commercial graphite-containing lubricant; the punch velocity was 1.7 mm/s. The final diameter of the extruded rod after the first pass was 8 mm, i.e., a reduction in wall thickness of $r_{T1} = 55\%$ was obtained, after the second pass 5 mm ($r_{T2} = 83\%$) and after the third pass 3 mm ($r_{T3} = 93\%$). The related ceramic core diameter, after the three extrusion passes, was 4, 2.5 and 1.5 mm, respectively. The explosively compacted billets were subsequently heated at 850°C in oxygen stream for 10 h and then slowly cooled at room temperature. The extruded rod, with diameter of 5 mm, after the second pass, was subjected to heat treatment in flowing oxygen at 920°C for 5 h followed by cooling at room temperature. Note that in both cases, the oxygen flow rates were up to 15 dm^3/h and the cooling/heating rates were equal to 300°C/h.

Punch load, P/punch travel, δ diagrams, as obtained from the autographic recorder of the hydraulic press, are presented in Figure 5.27. Similar to what has been reported for cold extrusion in (a) above, three distinct stages of the P-δ curve can be defined, namely I, II and III.

- Stage I is the transient or beginning stage of extrusion; where all the clearances between the die wall and the material were taken up and the plastic zone was developed with a small amount of extrusion occurring.
- Stage II corresponds to the steady-state stage, in which the load remains almost constant with a slight falling tendency; at this stage the composite material (silver and ceramic) is deformed plastically, whereas the friction force falls, reducing progressively the length of the slug remaining in the die.
- Stage III is probably associated with the beginning of the unsteady state, when the remaining slug is short; however, instability phenomena, leading mainly to possible hole-pipe formation obtained in cold extrusion, see (a) above, were not observed here.

The corresponding compression behavior of the explosively compacted billet is presented in a stress-strain plot shown in Figure 5.6, and it is consistent with the P-δ curves of Figure 5.27. As outlined in Section 5.2.3, during compression testing large amount of plastic deformation was imposed (an almost four times reduction of billet height was achieved accompanied by a characteristic barreling effect), without fracture or delamination, showing, therefore, a quasi-ductile overall material behavior, mainly attributed to the elevated-temperature processing ($\sim 0.6T_m$). On the other hand, the presence of silver in the ceramic core composition, say 10% Ag, may be responsible for the enhancement of formability and also of the density and fracture toughness, see

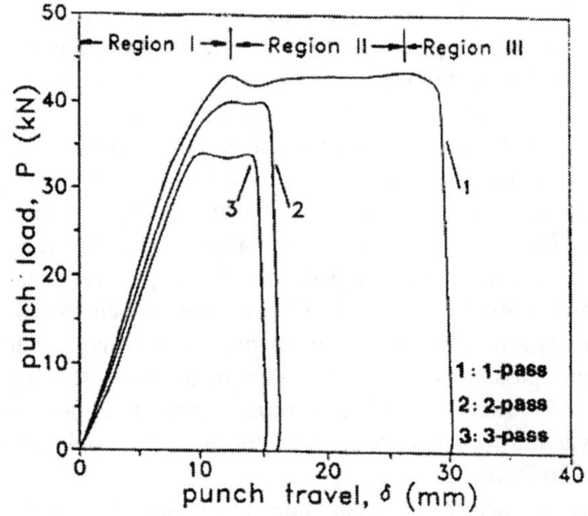

FIGURE 5.27. Punch load, P/punch travel, δ curves for the warm extrusion at 470°C of Ag/YBKCO billets; 1, 1st pass (r_T = 55%); 2, 2nd pass (r_T = 83%) and 3, 3rd pass (r_T = 93%).

References [34] and [35]. The compression curve occurs at the same temperature for a reduced silver content YBCO ceramic core composition shown in Figure 5.6 indicates a 10–20% lower ductility of the latter composite.

Microhardness measurements, performed on a diametral plane of the extruded rods, are shown in Figure 5.28; progressive "strenghtening" of the ceramic core was observed with increasing microhardness. The higher values of the microhardness are in good agreement with the ones corresponding to sintered YBCO ceramics, see References [36] and [37]. This observation is consistent with the microstructural refinement and the increase of bulk density after extrusion. Note also that the microhardness of the silver sheath is decreased because of the annealing process during heating.

Radial and longitudinal cracking, as well as metal/ceramic interface failure, may occur during cold extrusion of compacted and/or sintered ceramics, as outlined above. These failures can be avoided by increasing the formability of the ceramic core at the warm state. However, localized microcracks may be formed because of the development of tensile stress fields inside the billet, see Figure 5.29(a). The presence of flow lines, close to the metal/ceramic interface [see Figure 5.29(b)] often called *chevron cracks* or *internal cracks* or *shear fractures,* is due to the development of high secondary tensile stresses during extrusion [38]. This phenomenon, is more profound in the present case, due to the high thermal gradient developed near the YBKCO/Ag interface; the higher thermal conductivity of silver, compared with the ceramic, leads to

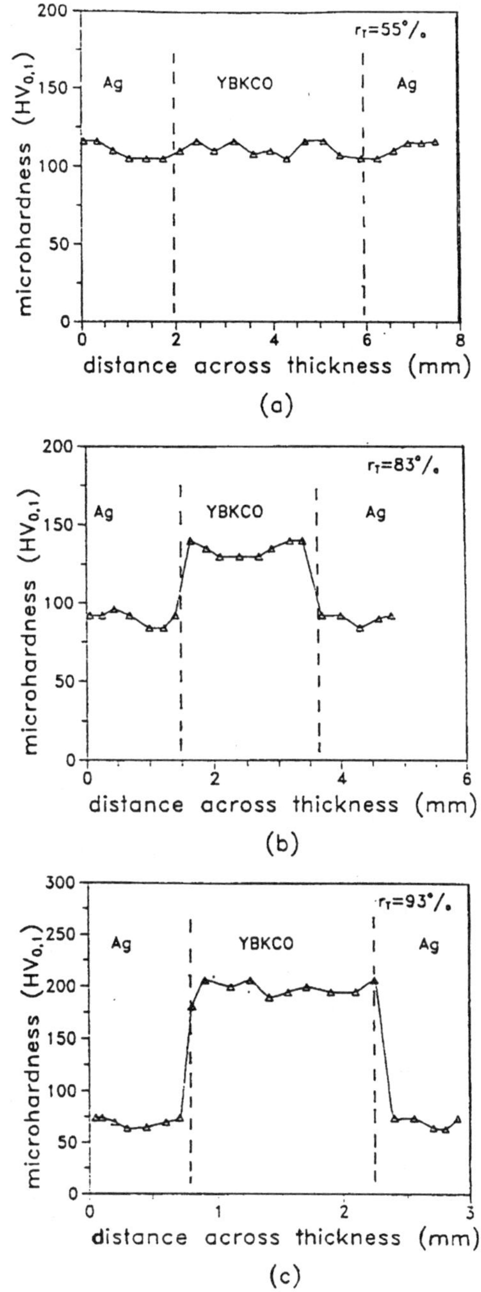

FIGURE 5.28. Microhardness profiles in a cross section of a warm extruded (470°C) Ag/YBCO ceramic composite rod after the (a) 1st extrusion pass (r_T = 55%); (b) 2nd extrusion pass (r_T = 83%); (c) 3rd extrusion pass (r_T = 93%).

rapid heat release through the metal sheath, creating, therefore, a chilled-non-deforming-zone. The evolution of the microstructure of the ceramic core during extrusion, as presented in Figure 5.30, indicates progressive reduction of the average grain size and subsequent increase of the bulk density. Minor secondary phase inclusions and limited microcracking are shown in Figure 5.29(a). The evolution of phase content and the critical temperature, T_c during fabrication, provided by XRD and dc-magnetic susceptibility measurements, respectively, are shown in Figures 5.31(a) and (b). The onset of the T_c, measured for the various passes of the progressively extruded HTS was equal to 93, 91 and 90 K, for the 1st, 2nd and 3rd pass, respectively, whereas the corresponding ΔT_c equals 3 K.

The extruded rod, of 5-mm diameter and 50-mm length obtained after the 2nd extrusion pass, was placed as an HTS conductor in a synchronous superconducting generator, after appropriate heat treatment. The T_c of the heat-treated component was 92 K, and the J_c was 200 A cm^{-2} at 77 K. The construction, design and performance of this machine are presented in Chapter 9.

PROFILE ROLLING OF METAL/YBCO WIRES

The explosive isodynamic compaction technique, schematically shown in Figure 5.32, was used for the fabrication of aluminium sheathed/$YBa_2Cu_3O_7$ billets for producing superconducting wire by subsequent profile rolling, see Reference [12]. After compaction, the billet [see Figure 5.32(b)], was rolled in a number of passes without intermediate annealing in a two-high rolling mill with rolls properly profiled to finally produce wire of a square cross section of 2.2-mm side length with a core of 1.3-mm diameter consisting of the $YBa_2Cu_3O_7$ superconducting compact; see Figure 5.32(c) and (d).

A transverse section of the fabricated wire, shown in Figure 5.32(d), reveals the fabrication of a "sound" component without defects and/or structural discontinuities and with good bonding characteristics between the superconducting core and the metallic aluminium sheath of the wire allowing for a proper electrical current transmission.

Longitudinal sections of the produced wire are shown in Figures 5.32(a) and 5.33(a). Strong bonding was obtained between the aluminium sheath and the superconducting core as indicated in Figure 5.32(b). Micrographs showing the shape and structure of the microstructural inclusions in the superconducting core, formed during compaction, after rolling, are presented in Figure 5.33(c) and (d). Note that after rolling, a separation at the interface between the inclusions and the matrix in the form of an arrow-shaped void occurred, see Figure 5.33(c); the latter defect is similar to the arrow-shaped defect developed because of excessive tension during metal extrusion and wire-drawing operations [2]. In some cases, rolling may also cause fragmentation of the inclusions, resulting in the formation of monocrystals with a very fine struc-

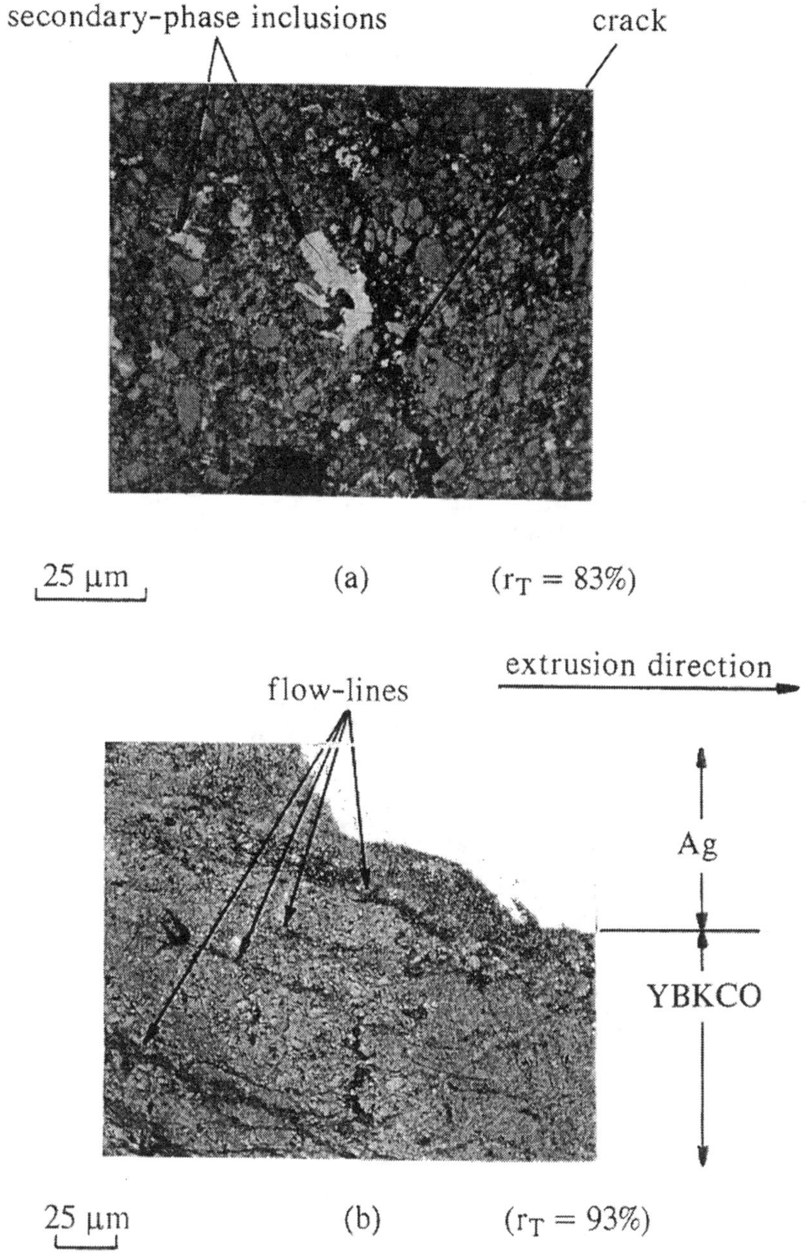

FIGURE 5.29. Optical micrographs showing various defects observed in warm extruded (470°C) Ag/YBCO rods after a total reduction: (a) r_T = 83%; (b) r_T = 93%.

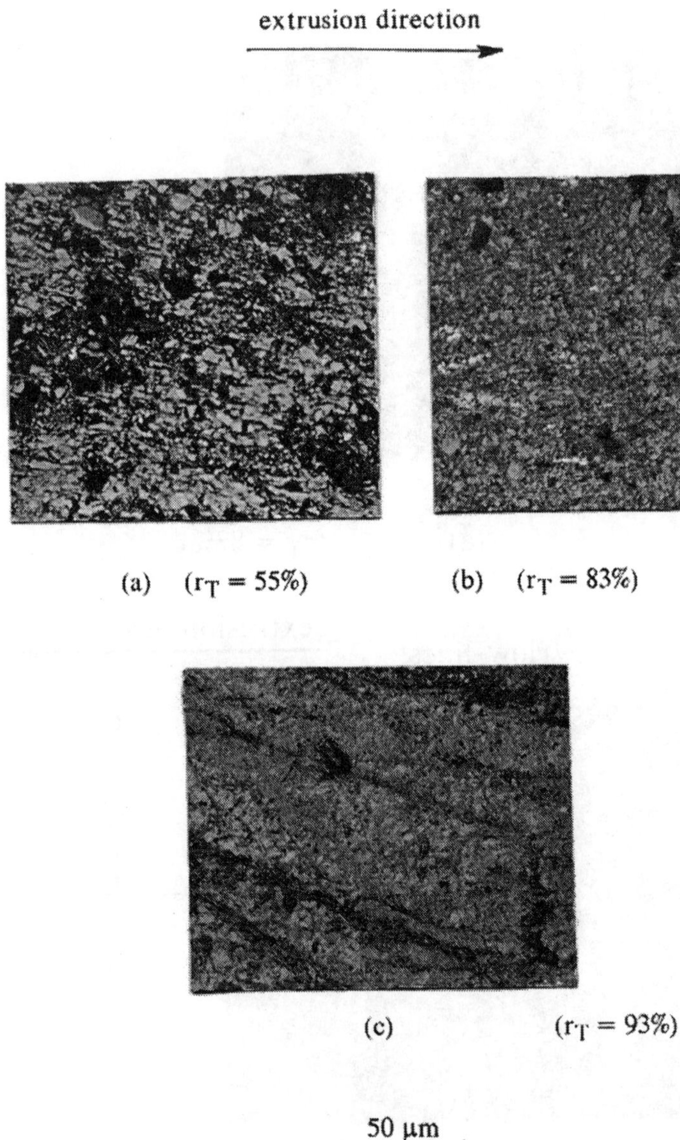

FIGURE 5.30. Optical micrographs showing the microstructure of the ceramic core of warm extruded (470°C) Ag/YBCO rods after the (a) 1st extrusion pass (r_T = 55%); (b) 2nd extrusion pass (r_T = 83%); (c) 3rd extrusion pass (r_T = 93%).

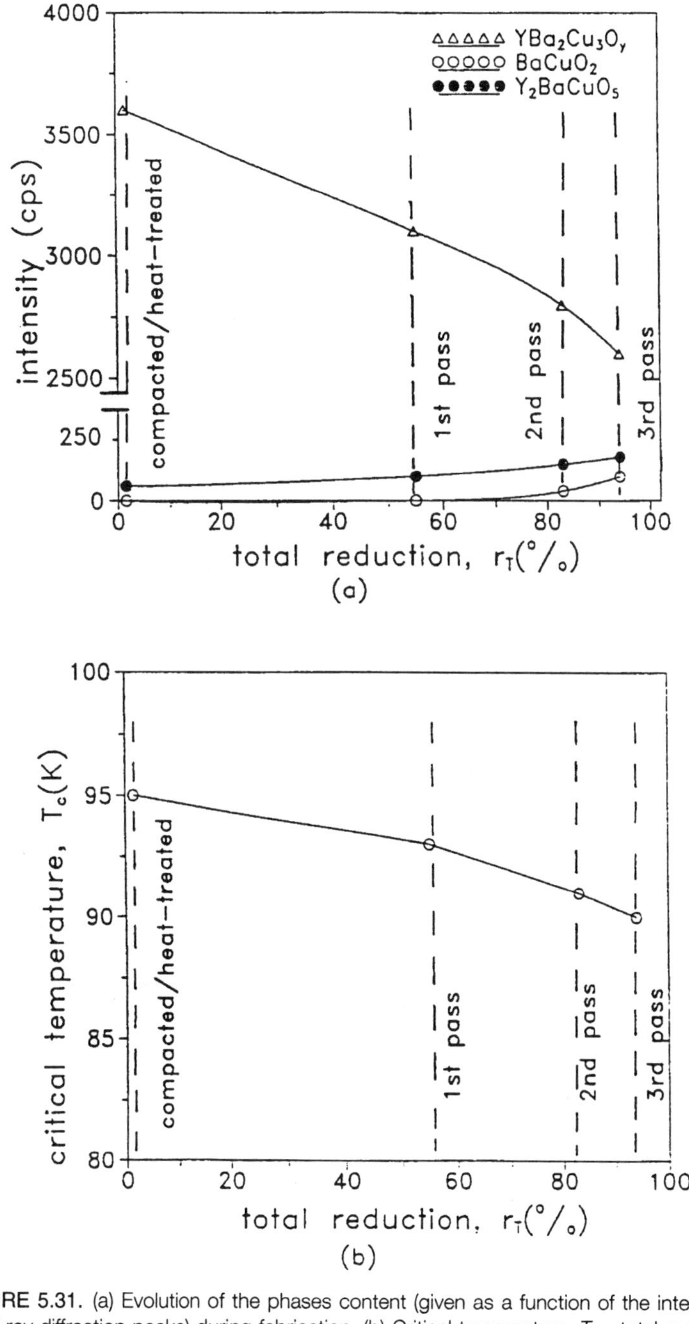

FIGURE 5.31. (a) Evolution of the phases content (given as a function of the intensity of the X-ray diffraction peaks) during fabrication. (b) Critical temperature, T_c - total reduction, r_T diagram for warm extruded (470°C) Ag/YBCO rods.

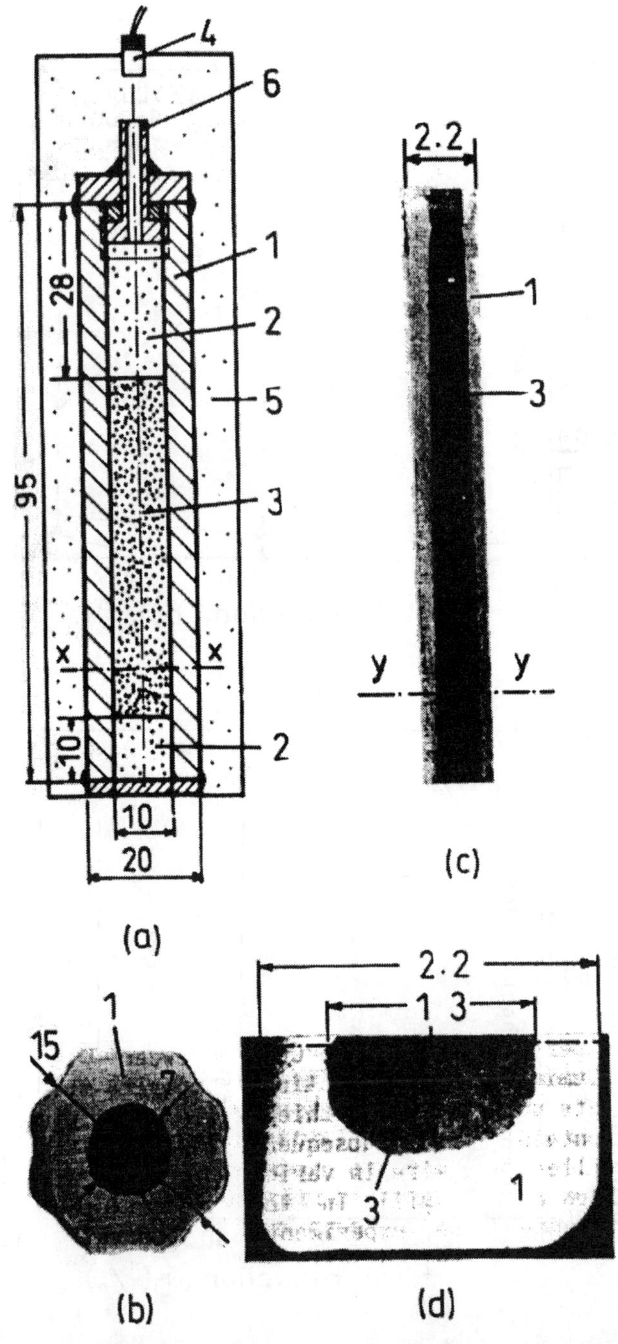

FIGURE 5.32. (a) Experimental explosive compaction setup; (b) a transverse section x-x of the compacted aluminium/YBCO billet; (c) a longitudinal section of the rolled wire; (d) a transverse section y-y of the rolled wire 1, Aluminium powder container; 2, MgO powder; 3, $YBa_2Cu_3O_7$ powder; 4, detonator; 5, high explosive; 6, vacuum tube (gas pressure 10^{-2} bar).

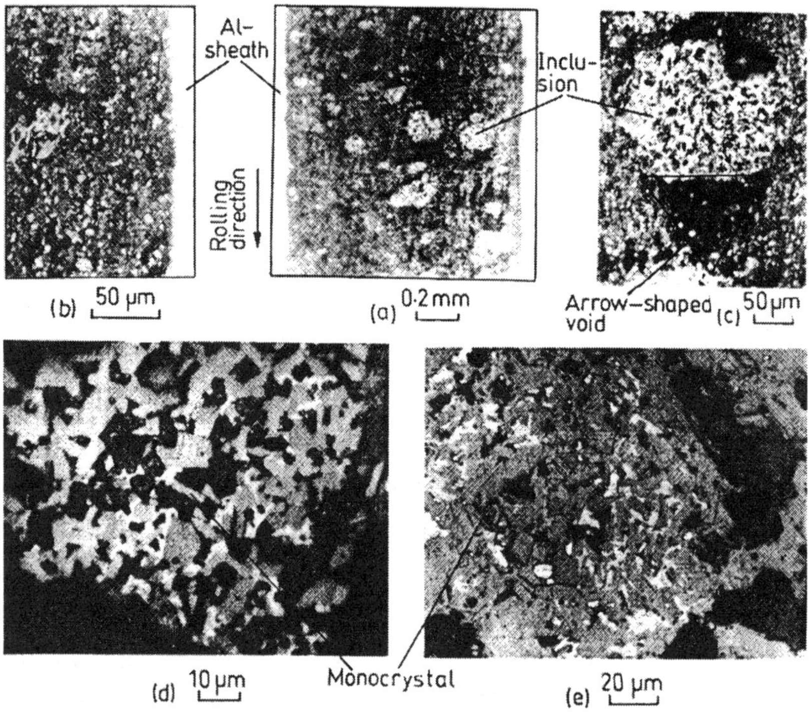

FIGURE 5.33. Optical micrographs showing the microstructural changes in a longitudinal section of an Al/YBCO profile rolled wire. (a) Shape and structure of the inclusions formed during compaction after rolling; (b) bonding at the interface of the aluminium sheath with the superconducting core; (c) arrow-shaped void formation from the separation of the inclusions with the matrix due to rolling; (d) fragmentation of the inclusions and formation of monocrystals; (e) microstructure of a sintered $YBa_2C_3O_7$ disk.

ture as indicated in Figure 5.33(d), but see also Figure 5.33(b). Note also that the microstructure of the superconducting core shown in Figure 5.33(d) is similar to the microstructure obtained after sintering of the explosively compacted $YBa_2Cu_3O_7$ disk shown in Figure 5.33 and reported in Reference [39], exhibiting strong Meissner effect and good superconducting properties.

From the magnetization measurements, performed on the longitudinal section of the wire shown in Figure 5.34, it is revealed that a critical temperature $T_c = 93.5$ K was obtained for the wire superconducting core, which is almost the same of that of the "green" compact. Therefore, it may be concluded, that the obtained superconductivity of the "green" compact is maintained after subsequent processing.

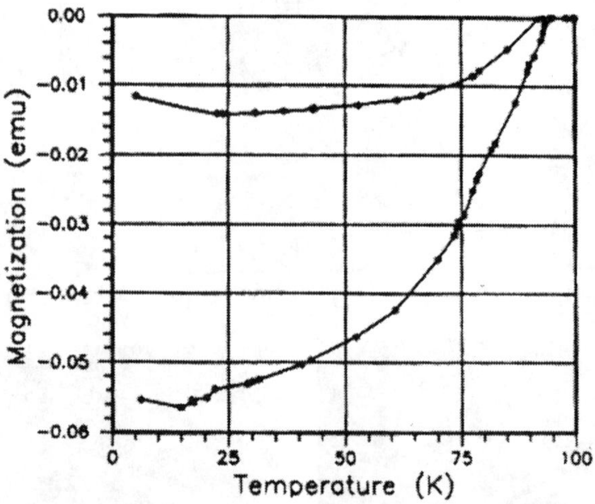

FIGURE 5.34. Magnetic susceptibility/temperature curve for profile rolled Al/YBCO wire.

5.6 FABRICATION OF FORGINGS

Mechanical aligning or texturing by pressing of superconducting powders can be applied as a secondary treatment for the fabrication of forgings of various geometries from already synthesized superconducting compounds. The concept of this technique is based on the fact that currents of very high density, up to 3000 A/cm^2, were observed on a thin surface layer of a press-forged bulk billet, which are attributed to the surface grain anisotropy. The mechanical texturing process leads to the production of bulk samples with their *c*-axis aligned parallel to the pressing direction.

The pressing operation can be realized in two stages, see Reference [40]:

- isostatic compression of the powder into homogeneous bars
- uniaxial forging with an overloading compressive stress under another isostatic confinement pressure regimen

The pressures of both cases varied between 0.1 and 1 GPa. The overloading uniaxial compressive stress is supposed to promote grain-alignment. The isostatic confinement leads to enhanced ductility of the material, whereas it behaves as a brittle solid under atmospheric pressure. The regions of different morphology due to inhomogeneous deformation, similar to the forging of metals, see Reference [1], is shown in Figure 5.35. Three distinct zones are designated:

- rigid or friction zone (I) due to the friction between the composite billet with anvil surfaces

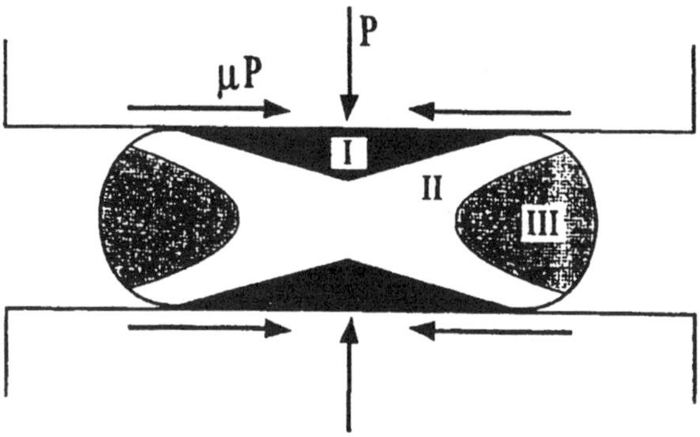

FIGURE 5.35. Deformation patterns of a superconductive compacted billet during mechanical texturing. I, high-friction dead zone; II, plastic zone; III, viscoplastic zone of intense cracking.

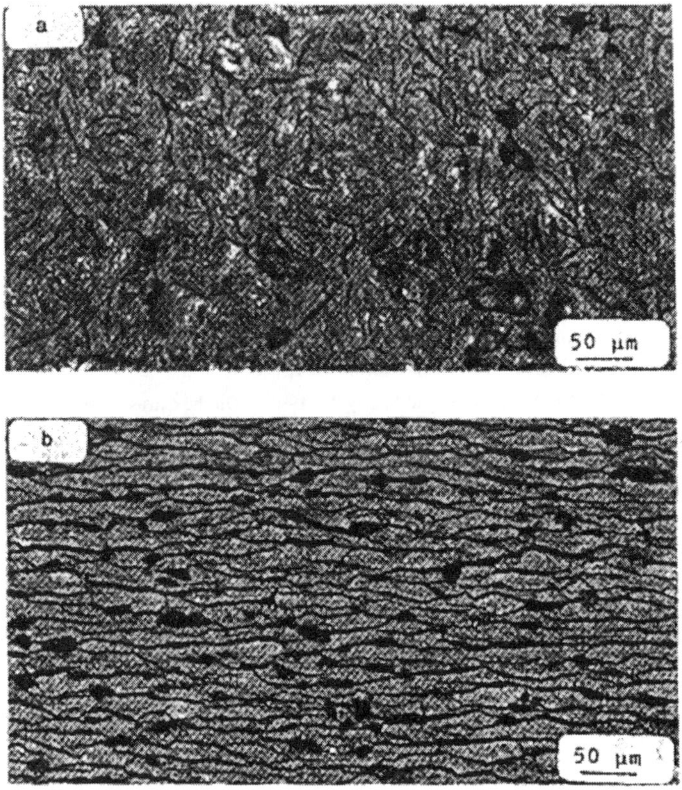

FIGURE 5.36. Optical micrographs showing the microstructure of a BSCCO billet: (a) precompacted and (b) mechanically aligned (black grains correspond to the Ca_2PbO_4 phase (from Reference [40]).

- plastic zone (II), where plastic deformation is confined along the shear planes or planes of discontinuities
- viscoplastic zone (III) of the compounds, where intense cracking occurs

The microstructure of BSCCO ceramic before and after mechanical alignment is shown in Figure 5.36(a) and (b), respectively [40]. A final sintering stage is involved for further densification.

5.7 REFERENCES

1. Johnson W. and Mamalis A. G. (1977), *Engineering Plasticity: Theory of Metal Forming Processes*. Springer Verlag, New York.
2. Johnson W. and Mamalis A. G. (1977), "A survey of some physical defects arising in metal working processes," *Proc. 17th Int. MTDR Conference,* Birmingham, UK, Mc Millan Press, London, 607.
3. Korzekwa D. A., Bingert J. F., Podtburg E. J. and Miles P. (1994), "Deformation processing of wires and tapes using the oxide-powder-in-tube method," *Appl. Supercond.* 2, 261.
4. Sakai T., Utsunomiya H., SaitoY., Hanamachi T. and Shinkawa M. (1997), "Effect of intermediate rolling condition on critical current density of silver-sheathed Bi(2223) superconducting tapes," *Physica C* 277, 189.
5. Mamalis A. G., Szalay A., Pantelis D. and Pantazopoulos G. (1996), "Net-shape manufacturing of silver-sheathed high-T_c superconductive ceramic (Y-Ba-K-Cu-O) by explosive compaction/cladding and rolling," *J. Mater. Proc. Technol.* 57, 112.
6. Mamalis A. G., Szalay A., Pantelis D. and Pantazopoulos G. (1995), "Fabrication of thick-layered superconductive ceramic (Bi-Pb-Sr-Ca-Cu-O)/metal composite strips by explosive cladding and rolling," *J. Mater. Proc. Technol.* 51, 255.
7. Mamalis A. G., Gioftsidis G. N. and Szalay A. (1992), "On the extrusion silver sheathed superconducting billets fabricated by explosive compaction of $YBa_2Cu_3O_7$," *J. Mater. Proc. Technol.* 30, 297.
8. Mamalis A. G., Szalay A., Pantelis D. I., Pantazopoulos G., Kotsis I. and Enisz M. (1996), "Fabrication of multilayered steel/superconductive ceramic (Y-Ba-K-Cu-O)/silver rods by explosive powder compaction and extrusion," *J. Mater. Proc. Technol.* 57, 155.
9. Mamalis A. G., Szalay A., Pantelis D. and Pantazopoulos G. (1995), "Net-shape manufacturing of silver/superconductive ceramic/silver rods by explosive compaction and warm extrusion," *Proc. EXPLOMET'95 Conference,* El Paso, Texas, August 1995; "Metallurgical and Material Applications of Shock-Wave and High-Strain-Rate Phenomena," J. L. E. Murr, K. P. Staudhammer and M. A. Meyers, eds., pp. 747–754.
10. Mamalis A. G., Pantazopoulos G., Szalay A., Kotsis I., Vajda I. and Manolakos D. E. (1996), "Multiple-pass warm extrusion of explosively compacted ceramic superconductive (Y-Ba-K-Cu-O)/metal billets," *Appl. Supercond.* 4, 213.
11. Mamalis A. G., Kotsis I., Vajda I., Szalay A. and Pantazopoulos G. (1996), "High-T_c ceramic superconductors for rotating electrical machines: from fabrication to application," *Proc. 10th Anniversary HTS Workshop on Physics, Materials and Applications*. World Scientific Publishers, Singapore, 631.

12. Mamalis A. G., Gioftsidis G. N. and Szalay A. (1990), "Fabrication of superconducting wire by explosive compaction and profile rolling," *Proc. 3rd ICTP Conference, Advanced Technology of Plasticity 1990,* Kyoto, Japan, 889.

13. Mamalis A. G., Vaxevanidis N. M. and Pantelis D. I. (1993), "On the rolling of bimetallic explosively cladded plates," *Proc. 4th ICTP Conference, Advanced Technology of Plasticity 1992,* 126.

14. Mamalis A. G., Vaxevanidis N. M. and Szalay A. (1996), "Cold rolling of multi-layer explosively cladded strips," *Int. J. Mach. Tools Manufacture* 36, 1033.

15. Wusatowski Z. (1969), *Fundamentals of Rolling,* Pergamon Press, Oxford, PA.

16. Mamalis A. G., Szalay A., Vaxevanidis N. M. and Manolakos D. E. (1998), "Fabrication of bimetallic rods by explosive cladding and warm-extrusion," *J. Mat. Proc. Technol.* 83, 48.

17. Han Z. and Freltoft T. (1994), "The mechanical deformation process for preparing Ag-sheathed BiSrCaCuO superconducting tapes," *Appl. Supercond.* 2, 201.

18. Lim H. J., Byrne J. G., Chae M. S. and Maple M. B. (1997), "Processing of BSCCO 2212 and 2223 superconductor compounds," *Mater. Manuf. Proc.* 12, 261.

19. Haugan T., Pitsakis M., Ye J., Wong F., Patel S. and Shaw D. T. (1995), "Fabrication of 90-m length $Bi_2Sr_2CaCu_2O_{8+x}$ tapes," *Appl. Supercond.* 3, 85.

20. Wu C. T., Goretta K. C. and Poeppel R. B. (1993), "Effects of processing paremeters on critical current density of Ag-clad $Bi_2Sr_2CaCu_2O_x$ tapes," *Appl. Supercond.* 1, 33.

21. Kumakura H., Togano K., Maeda H. and Mimura M. (1990), "Bi(Pb)-Sr-Ca-Cu-O superconducting composite tapes prepared by the powder method using an Ag sheath," *J. Appl. Phys.* 67, 3443.

22. Perin A., Grasso G., Daumling M., Hensel B., Walker E. and Flukiger R. (1993), "Hot rolling of silver-sheathed Bi(2223) tapes," *Physica C* 216, 339.

23. Guo J., Lewis J. A., Goretta K. C. and Schwartz J. (1995), "Properties and chemical stability of hot-rolled Ag(7 at. % Cu)-sheathed $Bi_2Sr_2Ca_{0.64}Cu_{1.64}O_x$ powder-in-tube tapes," *J. Appl. Phys.* 78, 4596.

24. Bernik S., Horvat M. and Kolar D. (1989), "Microstructural investigations of PbO modified Bi-Sr-Ca-Cu-O system," *Proc. ICMAS' 89,* Paris, France.

25. Shi D. (1989), "Formation of 100 K superconducting phase via amorphous state in the Bi-Sr-Ca-Cu-O system," *Phys. Rev. B* 39, 9091.

26. Ponnusamy D. and Ravi-Chandar K. (1993), "Extrusion of superconducting wires of $YBa_2Cu_3O_{7-x}$," *J. Mater. Res.* 8, 268.

27. Ponnusamy D., Coulter Y., Daugherty M., Ravi-Chandar K. and Salama K. (1996), "Processing of BSCCO superconducting rods for current leads," *Appl. Supercond.* 4, 247.

28. Samanta S. K., Wei Chen I., Wu X. and Tien T. Y. (1988), "Manufacturing of high-T_c superconducting ceramic wires by hot-extrusion," *Annals of CIRP* 37(1), 259.

29. Samanta S. K. (1990), "Processing and superconducting properties of $YBa_2Cu_3O_{7-x}$ + Ag_2O microcomposite for fabrication of high-T_c flexible wire," *Annals of CIRP* 39(1), 291.

30. Salama K. and Lee D. F. (1994), "Progress in melt texturing of $YBa_2Cu_3O_x$ superconductor," *Supercond. Sci. Technol.* 7, 177.

31. Watson D. R., Chen M. and Evetts J. E. (1995), "The fabrication of composite reaction textured $Bi_2Sr_3CaCu_2O_{8+\delta}$ superconductors," *Supercond. Sci. Technol.* 8, 311.

32. Woolf L. D., Raggio W. A., Elsner F. E., Fisher M. V., Stephens R. B., Figueroa T. L., Shearer

C. H., Rose J. D., Schaubel K. M., Olstad R. A., Ohkawa T., Duggan D. M., DiMartino M. and Fagaly R. L. (1991), "Continuous fabrication of high-temperature superconductor coated metal fiber and multifilamentary wire," *Appl. Phys. Lett.* 58, 534.

33. Selvamanickam V., Mironova M. and Salama K. (1993), "Enhancement of critical current density in $YBa_2Cu_3O_x$ superconductor by mechanical deformation," *J. Mater. Res.* 8, 249.

34. Matsumuro A., Kasumi K., Mizutami U. and Senod M. (1991), "Superconducting and mechanical properties of YBCO/Ag composites fabricated at high pressures up to 5.4 Gpa," *J. Mater. Sci.* 26, 737.

35. Singh J. B., Joo J., Singh D., Warzynski T. and Poeppel R. B. (1993), "Effects of silver additions on thermal shock and delayed failure of $YBa_2Cu_3O_{7-\delta}$," *J. Mater. Res.* 8, 1226.

36. Tuan W. H. and Tien T. C. (1994), "Mechanical performance of $YBa_2Cu_3O_{7-x}$-Ag composites," *J. Mater. Sci. Lett.* 13, 1019.

37. Ling H. C. and Yan M. F. (1988), "Microhardness measurements on dopant modified superconducting $YBa_2Cu_3O_7$ ceramics," *J. Appl. Phys.* 64, 1307.

38. Mamalis A. G., Kandeil A., de Mahlerbe M. C. and Johnson W. (1981), "Defects in cold-hydrostatically-extruded aluminium, iron and nickel-base compacts," *J. Mech. Work. Technol.* 4, 327.

39. Mamalis A. G., Gioftsidis G. N., Szalay A. and Boday O. (1989), "The shock-wave compaction of high temperature superconducting powders into cylindrical components," *Annals of CIRP* 38(1), 297.

40. Nhien S., Langlois P., Massat H., Wang J., Desgardin G., Lepropre M. and Provost J. (1994), "Bulk texturing of prereacted Bi/Pb(2223) under triaxial stresses at room temperature," *Physica C* 235–240, 3403.

CHAPTER 6

Fabrication of HTS Films

6.1 GENERAL

The development and optimization of manufacturing processes of bulk high-temperature superconducting samples are very important for the fabrication of components for electrical/electronic applications, such as conductors for high-power rotating machines, wires for current transport, coils, levitated bearings, etc. However, the production of HTS films are required in many (micro)electronic applications, ranging from strip-lines and hybrid interconnects to microwave antennas and from magnetic shielding devices to high-speed computer circuitry. These HTS films, needed for today's and tomorrow's electronic technology, should possess a variety of special properties, depending on the specific application, and a thickness varying from a few nanometers to hundreds of micrometer.

The processes used for the fabrication of HTS films can be divided into two main groups for the thickness of the film, namely:

- HTS thin-film processes
- HTS thick-film processes

6.2 HTS THIN-FILM PROCESSING

The most widely used processes in the electronic industry for fabricating HTS thin-films are:

- sputtering
- electron beam evaporation

- ion-plating
- laser ablation
- chemical vapor deposition (CVD)

The HTS thin-film coatings are produced by the *atom-transfer mechanism,* i.e., the HTS film is built up on the substrate by successive stacking of atomic layers. The HTS films are very dense and possess crystallographic order and preferred orientation/texturing during the epitaxial growth of the film on the substrate. For the YBCO ceramics the *c*-axis is perpendicular to the plane of the film and, therefore, the current flow is along the ab-planes, where the resistance is minimal. Texturing can be accomplished by selecting an appropriate single crystal or even a specially oriented substrate. Crystallographically oriented MgO, $SrTiO_3$ and $LaAlO_3$ compounds possess excellent lattice match with the Y-Ba-Cu-O material, leading to strong texturing and very high critical curent densities. Many electronic applications, ranging from computer memory devices to microwave cavities, require HTS thin films with very low surface resistance and high critical current density, J_c.

6.2.1 Sputtering

The sputtering method belongs to the *physical vapor deposition* (*PVD*) processes, which include also evaporation and ion-plating techniques. Sputtering techniques are based on the energetic collisions of gas ions produced by a glow discharge. The low-pressure gas atmosphere (usually Ar or Ar/O_2) is ionized during high-voltage electrical discharge, using a dc or an ac power supply, and the induced energetic gas cations bombard the target (cathode), see Figure 6.1. This cathode is made of the material to be deposited on the substrate, e.g., YBCO or BSCCO. Sputtering is not a very efficient technique, judging from the deposition rate. Single target sputtering leads to the pro-

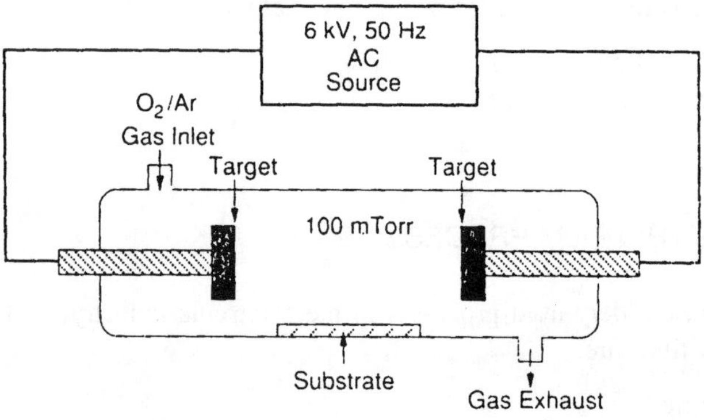

FIGURE 6.1. A schematic representation of a double target ac-sputtering system (from Reference [2]).

duction of highly stoichiometric 123 thin films, with thickness ranging from 1 to 1000 nm and possessing enhanced superconducting properties [1].

Dual target sputtering has been reported, where one target of yttrium was supplemented with another of Ba_2Cu_3 compound; the oxygen was added in a postannealing treatment [1]. BSCCO thin films can be produced by multiple target ac-sputtering on polycrystalline yttria-stabilized-zirconia substrate (YSZ), see Reference [2]. Postdeposition annealing treatment at 750–900°C has led to a T_c onset up to 115 K, whereas zero resistivity was observed at 73 K because of the presence of both 2223 and 2212 phases.

6.2.2 Electron Beam Evaporation

Evaporation techniques are the most simple PVD methods. The material to be evaporated is placed in a vacuum chamber near the substrate, vaporized by means of thermal energy provided by a heating source, whereas the atoms in the vapor phase are solidified on the substrate. Coatings of thickness ranging from 1 to 1000 nm can be obtained. A variety of methods can be used to heat the material; the most common technique is the electron beam heating among them. Vacuum in the range of 10^{-5}–10^{-6} Torr is generally used, allowing for the atoms to follow a line-of-sight trajectory to the target substrate without significantly lowering their energy by mutual collisions.

The electron beam evaporation chamber is shown schematically in Figure 6.2, see also Reference [3]. Because the evaporated material will not be uniformly deposited on the target, the target may be rotated during the process to improve the homogeneity of the film.

6.2.3 Ion-Plating

The ion-plating process is very similar to the electron beam evaporation technique; the only difference is that the evaporated material passes through a glow discharge on its trajectory to the substrate; this electrical discharge leads to ionization of a portion of atoms, while the substrate is biased to about 2 kV (negative potential) to ionize argon. This procedure results in cleaning the substrate surface via the energetic collisions of gas ions. The surface generated by the collisions, e.g., removed impurities, roughness change, etc., leads to the improvement of the adhesion of the coating. The deposition rate of the plating process is relatively low, leading to coatings of thickness varying between 10 and 10,000 nm.

6.2.4 Laser Ablation

Laser ablation constitutes one of the most important deposition techniques for the manufacture of HTS thin films. A typical configuration of an excimer

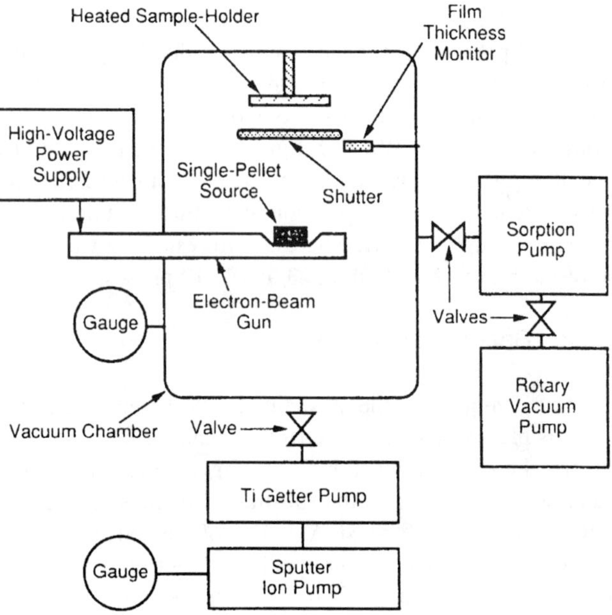

FIGURE 6.2. Experimental setup of an electron-beam evaporation chamber (from Reference [3]).

laser source (pulsed-laser) emitting characteristic radiation of 248 nm is shown in Figure 6.3, see References [4,5]. The photochemical interaction between the laser beam and the target leads to evaporation of surface atoms of the target, consisting of YBCO or BSCCO compounds. Successful films may be produced by using fluences of a few J/cm^2. Because of the short wavelength absorption, the ablated material and the ionized gas atoms are excited to fluorescence or chemiluminescence, creating a light-emitting plume. The spieces, transferred to the target, result in the formation of a highly stoichiometric film. A normal deposition rate is about 10 nm/min, but films of thickness ranging from 10–1000 nm can be obtained.

The substrates used must possess chemical inertness and similar thermal expansion coefficient with the superconducting films. The substrate material is preheated to about 750°C before the deposition process. Such substrates are MgO, glass, sapphire, YSZ and perovskite materials like $SrTiO_3$, $LaGaO_3$ and $LaAlO_3$. Postdeposition annealing in oxygen atmosphere is, in many cases, necessary to establish the desired oxygen stoichiometry in the film.

6.2.5 Chemical Vapor Deposition (CVD)

Chemical vapor deposition (*CVD*) is a chemical process for the production of a variety of engineering materials from hard TiN and nanocrystalline dia-

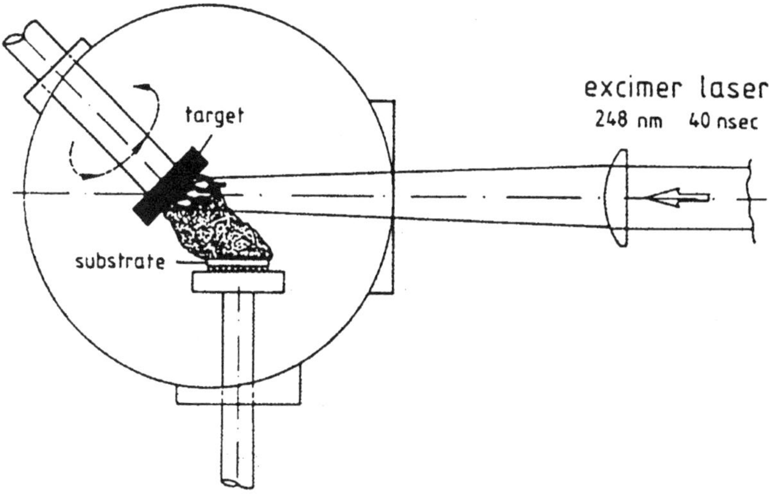

FIGURE 6.3. A typical excimer-laser ablation configuration (from Reference [5]).

mond to high-temperature superconducting coatings. The CVD process is a one-step deposition technique that includes the chemical synthesis and the coating of HTS material.

A special CVD reactor for the synthesis of BSCCO superconducting coatings is shown in Figure 6.4, see Reference [6]. It consists of an outer horizontal furnace, creating a temperature gradient along the reactor, whereas an opened quartz tube containing stoichiometric proportions of the reactants in chloride and iodide compounds is placed in the interior of the reactor. Carrier gas mixtures, containing He and O_2, H_2O, flow in the reactor and decompo-

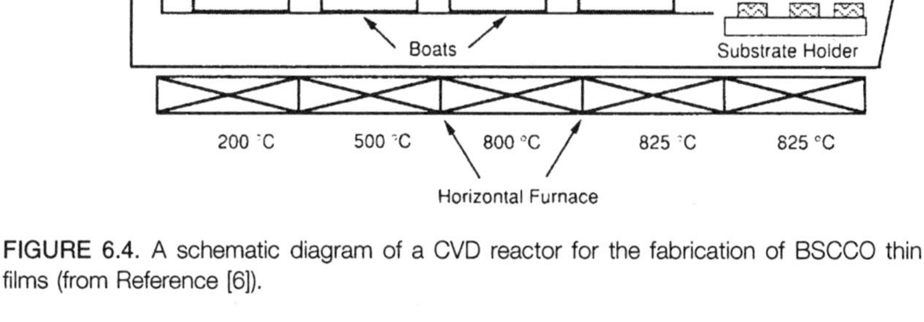

FIGURE 6.4. A schematic diagram of a CVD reactor for the fabrication of BSCCO thin films (from Reference [6]).

sition and reconstruction reactions take place, resulting in synthesizing a complicated CVD kinetic schema, leading, finally, to the formation of the superconducting ceramic deposits on the substrates. The advantage of the CVD technique is the use of cheap raw materials instead of relatively expensive superconducting precursors.

6.3 HTS THICK-FILM PROCESSING

The main processes used in the electrical/electronic industry for producing high-temperature superconductive thick films may be listed:

- screen-printing
- in situ melting
- electrophoresis
- powder plasma-spraying (arc-spraying, high-velocity oxy-fuel flame spraying (HVOF))

HTS coatings at the liquid or at the solid state are produced by using thick-film processes. In screen printing, various thermal cycles at the solid state are used, whereas the plasma-spraying and the related techniques are used for the manufacture of thick HTS coatings by the droplet-transfer mechanism. The atom-transfer mechanism, which is the main mechanism in thin-film processing, see Section 6.2 above, constitutes also the governing mechanism of the electrophoresis process for fabricating thick films. The thickness of the coating ranges from tens to hundreds of micrometers and exhibits a porous and granular microstructure. Microcracks and local coating debonding are often present in the film area. Coating with such a microstructure and defects (surface roughness, microcracking, granularity, lack of adhesion) are inproper for high-current applications. They can be used as components for low-J_c and magnetic applications including SQUIDs, IR detectors and magnetic shielding equipment.

6.3.1 Screen Printing

YBCO and BSCCO ceramic thick films from 20 to 200 μm can be produced by the screen-printing method. These thick films have several applications in interconnections of hybrid circuits, as well as microwave circuits, whereas their low-cost fabrication technique is beneficial in the microelectronics industry. BSCCO ceramic seems to be the dominant material in screen printing. The substrate materials can be metals, such as Ag, as well as ceramics, such as Al_2O_3 or YSZ [7,8]. The screen-printing process, for the fabrication of BSCCO thick films, consists of the following steps:

- synthesis of the 2212 phase powder
- mixing of the ceramic powder with various organics (solvent, binder, plasticizer) to form a paste
- screen printing of the paste on the substrate through a 300-mesh silk screen
- firing at low temperatures, e.g. 500–550°C, for the removal of the organics
- final heat treatment at higher temperatures, e.g., 750–880°C, for sintering the ceramic powder

6.3.2 In Situ Melting

The in situ melting technique is a relatively simple method for the production of BSCCO films with a thickness varying from 10 to 50 μm on alumina or magnesia substrates [9]. This technique is also known as the *melt-quenching* process and consists of the following stages:

- synthesis of the precursor high-T_c BSCCO powder
- grinding, refiring and sieving the resultant ceramic powder to very fine grain sizes (2–3 μm)
- distribution of the fine powder on the substrate
- melting process in air at 1000–1050°C followed by quenching
- heat-treatment at 850–870°C

The film produced after quenching exhibits glassy-like structure with semiconducting behavior. The final heat treatment stage is necessary, to transform the glass compound to the corresponding crystalline material with the desired superconducting properties.

6.3.3 Electrophoresis

Electrophoresis is the electrochemical deposition of HTS compounds on various substrates, e.g., Ag, Au, oxides, perovskites. The material of the YBCO or BSCCO compounds or the raw materials, such as oxides, to be deposited, forms a colloidal suspension or a solution based on organic solvents, such as a mixture of butanol-propanol [10,11]. The method is very efficient for the production of relatively thick coatings ranging from 10 to 50 μm. Postdeposition heat treatment is necessary for sintering and oxygenation of the produced ceramic coatings.

6.3.4 Plasma-Spraying

The plasma-spraying and related techniques have been extensively used for the fabrication of thick ceramic or intermetallic coatings, of thickness ranging from 0.1 to up to several mm, with enhanced surface properties, e.g., extreme hardness and wear resistance, thermal insulation, thermal shock resistance and corrosion resistance. Plasma coating is succesfully applied to various engineering components, such as aerospace materials, surgical implants, cutting tools and machine parts, that are subjected to severe friction and high-temperature loading conditions. The plasma-spray techniques provide the advantages of reaching high-coating density and superior mechanical behaviour over large surface areas and also the ability to cover nonuniform shapes. The versatility of the plasma-spraying techniques as well as the material-processing advantages, enable this method to be used in the fabrication of thick HTS coatings, e.g., 100–200 μm, for the microelectronics industry [12]. This plasma-spraying technique was recently used for the fabrication of high-T_c superconducting coatings of the Y-Ba-Cu-O compound, see References [13]–[18] and the Bi-Sr-Ca-Cu-O compound, see References [12], [19] and [20], on various substrates for advanced electrical/electronic applications, such as magnetic shielding devices, bolometers, microwave cavities, etc.

Several spraying techniques have been used for the fabrication of thick coatings:

- atmospheric plasma-spraying (APS)
- vacuum plasma-spraying (VPS)
- detonation gun
- combustion flame spraying
- hypervelocity oxy-fuel flame spraying (HVOF)

A typical schematic diagram of the plasma-spraying process is shown in Figure 6.5. During plasma-spraying, the material to be deposited is injected into a plasma torch in powder form. At the extreme plasma temperatures, 10,000–20,000 K, the powder melts, and the molten droplets are accelerated onto the substrate with a velocity up to several hundreds of meters per second. On impingement, the droplets splat and rapidly solidify, with cooling

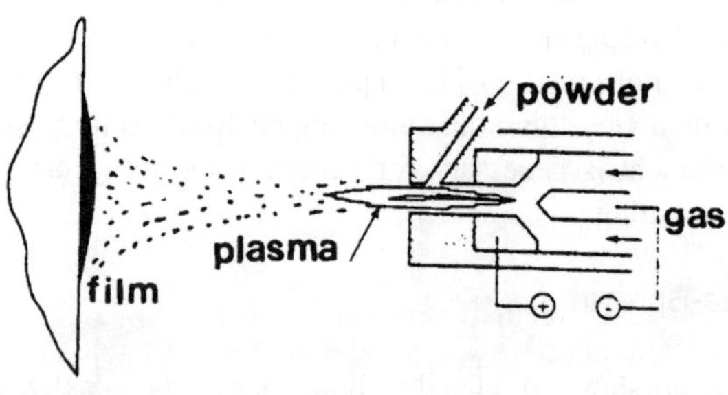

FIGURE 6.5. A schematic representation of the plasma-spraying process.

rates up to 10^5 K/s. The connection of splats creates a single plasma-coating layer, whereas the whole coating is built up by stacking of several layers.

The developed high temperature of the droplets, due to their collision with the substrate, may cause undesirable reactions or diffusion and contamination of the plasma coating. Inert substrates and intermediate layers, such as yttria-stabilized zirconia, acting as diffusion barriers, are proposed to obtain sound and "clean" plasma coatings, see Reference [12]. The rapid quenching, after the spraying process, does not permit the regrowth of the superconducting phase. The plasma coating, after the rapid solidification, consists of a mixture of oxides, produced by the thermal decomposition of the initial compound plus a glassy phase. In many cases, the tetragonal nonsuperconducting 123 phase for YBCO powders is apparent [13]. Postdeposition heat treatment is often necessary for the crystallization and regeneration of the superconducting phase.

6.3.5 Fabrication of Thick Coatings by Plasma-Spraying and Excimer Laser Post-treatment

The excimer laser processing can be used as a postdeposition surface treatment of plasma-sprayed thick superconducting ceramic coatings. Excimer laser irradiation, emitted in the UV range, provides new facilities for material processing and surface treatment, especially in the micrometer range [21]. The excimer laser material interaction leads to polishing of the surface by remelting and rapid resolidification or to very accurate material removal, described by the well-known photochemical ablation mechanism, generating, therefore, new microstructures due to the short UV wavelength, the high energy rate received per material volume and the excellent focusing properties of the laser beam. The numerous advantages of the excimer laser irradiation make this technique very attractive, mainly in the field of microelectronics-processing technology. Note also that thin superconducting films, of thickness less than 1 μm, on crystallographically oriented substrates, e.g., MgO, are fabricated by using excimer laser irradiation with the laser-ablation-deposition technique, see Section 6.2.4.

Postprocessing heat treatment in flowing oxygen is necessary to recover the superconducting properties of the laser-treated superconductive ceramic coating. Heating at high temperatures, up to 900°C, activates diffusion processes, leading, therefore, to the reduction of the porosity and to the enhancement of crystallinity.

Thick YBKCO superconducting coatings, up to 150–200 μm, on 50 × 50 mm and 3-mm thick silver plates, were fabricated by using the atmospheric plasma-spraying (APS) method as deposition technique, followed by excimer laser irradiation and subsequent heat treatment in oxidizing atmosphere, see

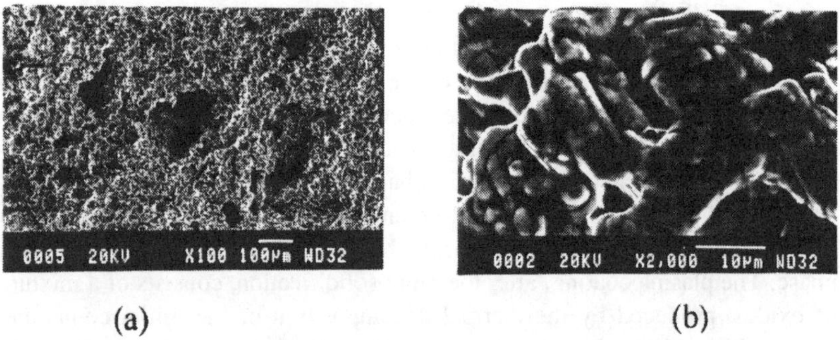

FIGURE 6.6. (a) Scanning electron macrograph of the pelletized superconducting powder. (b) Scanning electron micrograph showing the microstructure of the initial superconductive powder.

Reference [18]. The powder mixture used, with the nominal composition of $YBa_{2-x}K_xCu_3O_v$ (x = 0.01–0.10), was prepared by using stoichiometric quantities of Y_2O_3, $Ba(OH)_2 \cdot 8H_2O$, $KF \cdot 2H_2O$ and CuO. The procedure for producing particles, with size varying between 63 and 90 μm, is described in detail in Reference [22]; see also Section 3.4.1. The macro- and microstructures of the produced superconducting powder are shown in Figure 6.6, whereas the corresponding X-ray diffraction pattern revealing the presence of the orthorhombic superconducting phase is presented in Figure 6.7.

The plasma-spraying parameters used were as follows: arc current = 300 A; arc voltage = 50 V; argon flow-rate = 2 m^3/h; and nitrogen flow rate = 0.5 m^3/h. Laser treatment of large areas, using an excimer laser, operating at a wavelength of 248 nm (KrF) with a rectangular energy beam profile, was

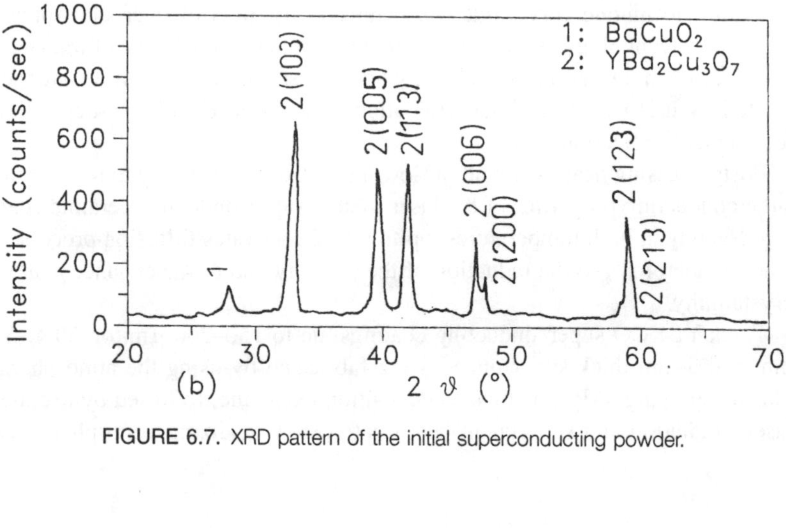

FIGURE 6.7. XRD pattern of the initial superconducting powder.

performed by 50% overlapping between successive single laser spots, realized by using a constant number of pulses up to 15. Two different energy densities per pulse were used, namely 6 and 8 J/cm^2, whereas the frequency varied between 10 and 50 Hz. Subsequently, postprocessing heat treatment in oxidizing atmosphere was performed in a programmable furnace at heating temperature of 900°C; the heating duration was 2 h, and the related heating and cooling rates were 300°C/h and 60°C/h, respectively, wheras the oxygen flow rate varied between 0.05 and 0.07 m^3/h.

Scanning electron micrographs of the plasma-sprayed superconductive ceramic revealed an island-like structure with weak links between the superconducting grains, see Figure 6.8(a) and (b).

Surface modifications, induced by subsequent excimer laser treatment, are shown in Figure 6.8(c)–(f). During the excimer laser material interaction the following processes take place successively:

- Remelting of a thin surface layer, less than a grain size, when low-beam energies are used, eliminating the initial grain boundaries of the as-sprayed coating, see Figure 6.8(c), or of deeper surface layers, of a few grain sizes, leading to the refinement of the near-surface microstructure and the formation of new grain boundaries, for higher beam energies; compare Figure 6.8(e) with Figure 6.8(a) and (c).
- Material evaporation for high laser beam energies
- Rapid solidification of the remaining molten material, resulting in the development of intense tensile stress fields leading to microcracking, see Figure 6.8(d) and (e).

A lamellar-type morphology, similar to that observed in many eutectic or eutectoid systems, resulted from localized ablation of species in the vicinity of grain or phase boundaries, is shown in Figure 6.8(f). This wavy surface pattern is very characteristic in excimer laser materials processing. The "ripple formation" is a complex result that depended on fluid motion and surface tension-associated phenomena. Similar observations, reported in Reference [23], are attributed to the "optical etching" effect.

Finally, SEM examinations of the heat-treated ceramic, reveal a sponge-like microstructure with strong grain links that resulted from intense particle boundary diffusion; see Figure 6.8(g) and (h).

In Table 6.1 are the EDAX analysis results, indicating the chemical composition of the surface and the near-subsurface region of the coatings; changes in stoichiometry of the ceramic coating after the various stages of the fabrication are revealed. Concerning the plasma-sprayed ceramic, the detected molar ratio confirms the decomposition of the initial superconducting compound to yttrium-based oxides. Improvement in stoichiometry was achieved after the excimer laser surface treatment, mainly because of the crystallization of the

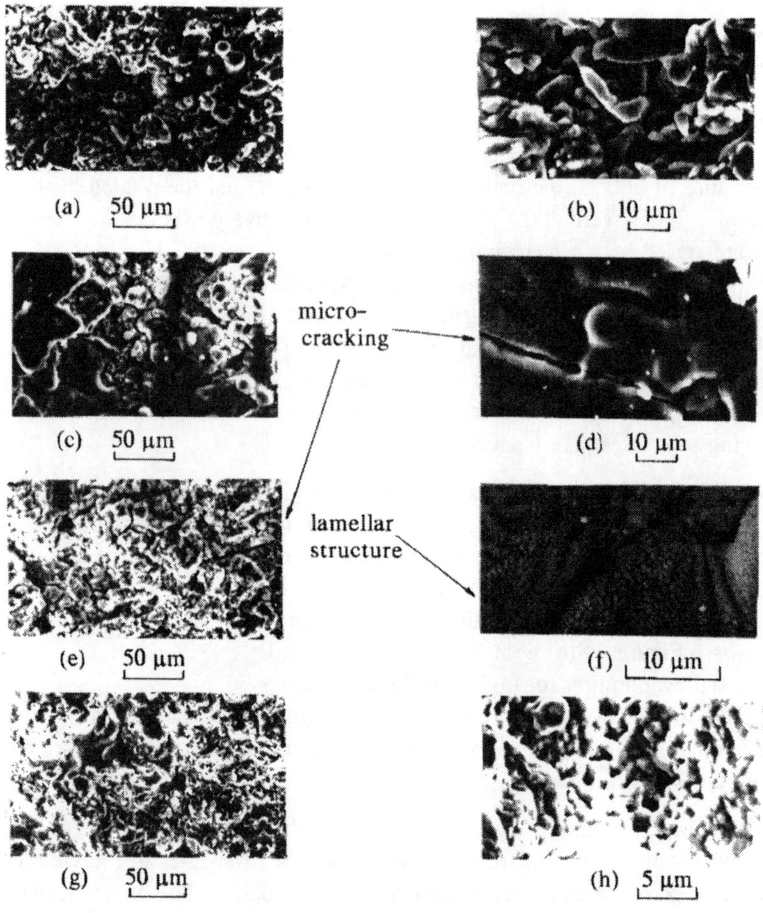

FIGURE 6.8. Scanning electron micrographs showing the microstructure of the YBKCO superconductive ceramic coating after the various stages of fabrication; (a) plasma-sprayed; (b) detail of (a); (c) laser treated with 6 J/cm^2 per pulse; (d) detail of (c); (e) laser treated with 8 J/c,^{2} per pulse; (f) detail of (e); (g) heat-treated; (h) detail of (g) (from Reference [18]).

TABLE 6.1. EDAX Surface Chemical Analysis for the Various Stages of Fabrication of YBKCO Coating.

Specimen	Y:Ba:Cu (molar ratio)
Initial powder	1 : 2.19 : 4.28
Plasma sprayed	1 : 1.34 : 1.11
Laser treated (E = 6 J/cm^2)	1 : 1.85 : 3.16
Laser treated (E = 8 J/cm^2)	1 : 1.75 : 2.81

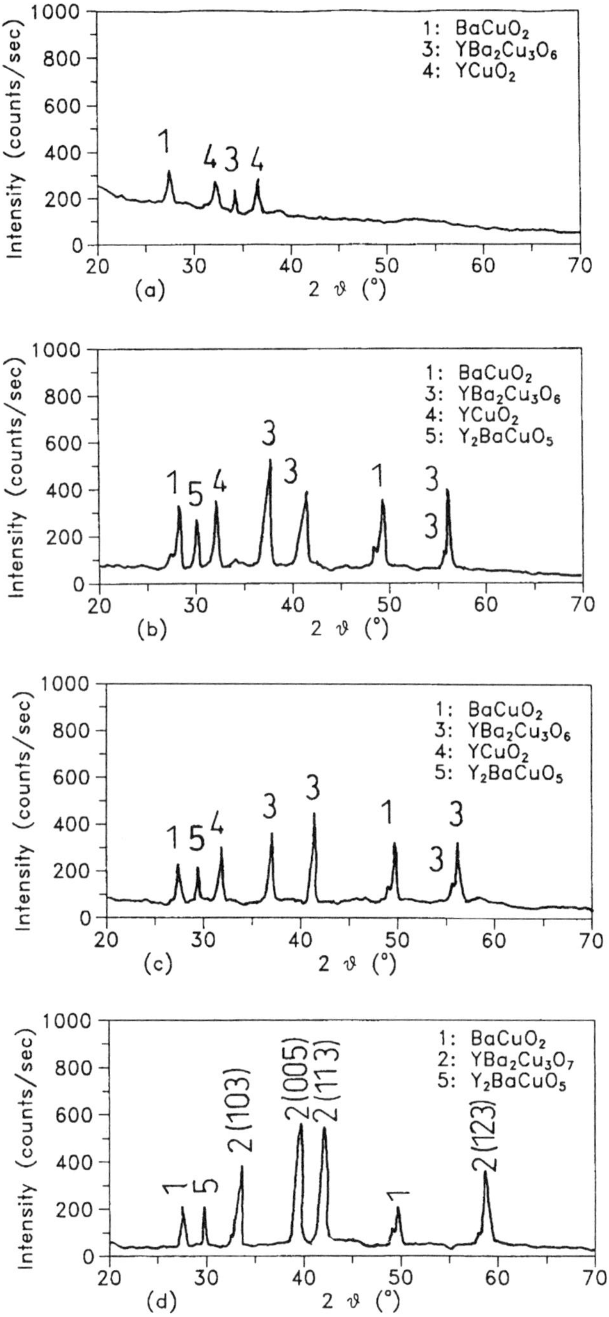

FIGURE 6.9. XRD patterns of the YBKCO ceramic coating after the various stages of the fabrication; (a) plasma-sprayed; (b) laser-treated with 6 J/cm^2 per pulse; (c) laser-treated with 8 J/cm^2 per pulse and (d) heat-treated (from Reference [18]).

1:2:3 phase in a great percentage, with the tendency to reach the stoichiometry of the initial powder.

XRD analysis performed on the surface of the plasma-sprayed coating did not show significant crystallographic reflection peaks, see Figure 6.9(a), indicating, therefore, not only the loss of the orthorhombic phase but also an amorphous-like structure that resulted from fusion and very rapid solidification induced by plasma; traces of the $BaCuO_2$, $YCuO_2$ compounds and the tetragonal non-superconducting $YBa_2Cu_3O_6$ phase were detected. Enhancement of crystallinity of the laser-irradiated coatings, due to the resulted lower solidification rates as compared with those induced by plasma-spraying, is revealed in Figure 6.9(b) and (c). However, the interaction between excimer laser and ceramic coating favors the crystallization kinetics of the tetragonal nonsuperconducting phase, resulting in a partial or even complete loss of superconductivity, and, therefore, an appropriate postprocessing heat treatment is required to reestablish the superconductivity of the ceramic deposits. In the latter case, lattice diffusion of oxygen atoms occurred, filling the oxygen vacancies and resulting in the formation of the orthorhombic superconducting phase, see Figure 6.9(d). Note also that the adhesion of the superconductive ceramic coatings with the silver substrate was improved after laser processing, indicating the applicability of the excimer laser processing to fabricate sound components.

The voltage current intensity behavior of the plasma-sprayed YBKCO ceramic, possessing a glassy-like structure, presented in Figure 6.10, reveals semiconductor-like properties, similar to the characteristic curve of Germanium-type transistor. Semiconducting behavior of plasma-sprayed high-T_c ceramics was also reported in Reference [14].

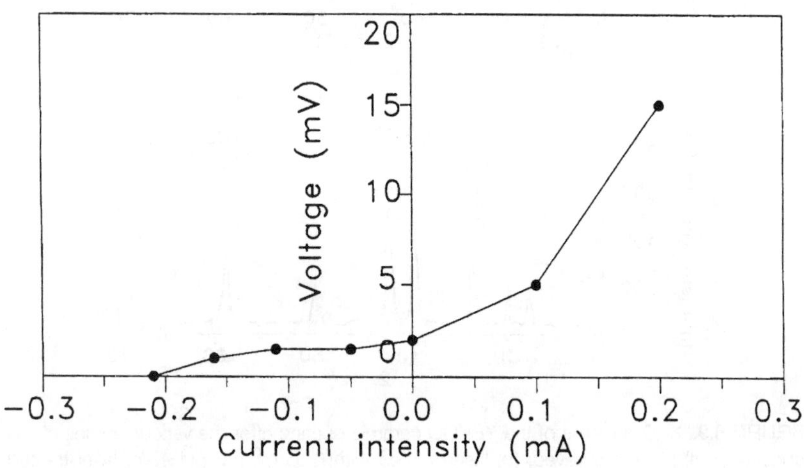

FIGURE 6.10. Voltage-current diagram for the plasma-sprayed YBKCO superconductive ceramic coating (from Reference [18]).

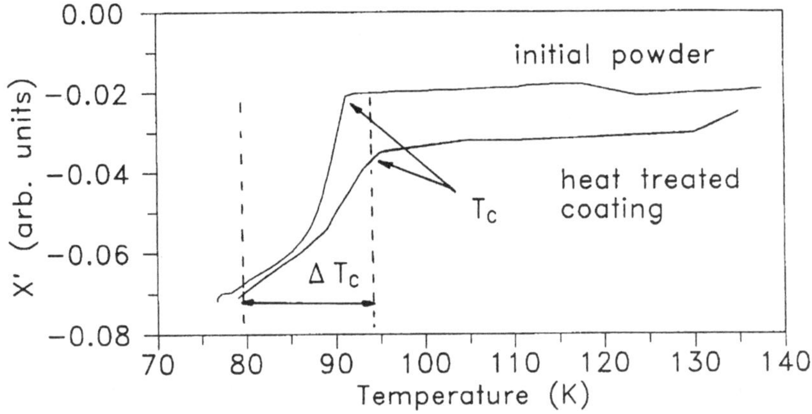

FIGURE 6.11. The ac-susceptibility (χ')-temperature diagram for the initial YBKCO powder and the heat-treated superconductive ceramic coating (from Reference [18]).

The real part of the ac-magnetic susceptibility (χ'-component), corresponding to the d*M*/d*H* (*M,* magnetization; *H,* magnetic field strength), as a function of temperature is presented in Figure 6.11. The normal to superconducting state transition of the postheat-treated coating, detected by the diamagnetic Meissner shielding signal, was observed at an onset temperature up to 92 K, whereas the corresponding transition width, ΔT_c was around 15 K, for both energy densities per pulse examined. The lower transition height and larger width of the corresponding transition curve indicate the presence of lattice defects, greater dislocation density and lower volume fraction of the superconductive phase compared with that referred to the initial powder.

6.4 REFERENCES

1. Makous J. L., Maritato L., Falco C. M., Cronin J. P., Rajendran G. P., Uhlmann E. V. and Uhlmann D. R. (1987), "Superconducting and structural properties of sputtered thin films of $YBa_2Cu_3O_{7-x}$," *Appl. Phys. Lett.* 51, 2164.
2. Koinuma H., Kawasaki M., Nagata S., Takeuchi K. and Fueki K. (1988), "Preparation of high-T_c Bi-Sr-Ca-Cu-O superconducting thin films by ac sputtering," *Jap. J. Appl. Phys. Lett.* 27, 376.
3. Terasaki I., Nakayama Y., Uchinokura K., Maeda A., Hasegawa T. and Tanaka S. (1988), "Superconducting films of $YBa_2Cu_3O_x$ and Bi-Sr-Ca-Cu-O fabricated by electron-beam deposition with a single source," *Jap. J. Appl. Phys. Lett.* 27, 1480.
4. Venkatesan T., Wu X. D., Inam A., Jeon Y., Croft M., Chase E. W., Chang C. C., Wachtman J. B., Odom R. W., di Brozolo F. R. and Magee C. A. (1988), "Nature of the pulsed laser process for the deposition of high T_c superconducting thin films," *Appl. Phys. Lett.* 53, 1431.
5. Fröhlingsdorf J., Zander W., Stritzker B., Feile R. and Leiderer P. (1989), "Macroscopic persistent currents in laser deposited $YBa_2Cu_3O_7$ films," *Physica C* 159, 513.

6. Ihara M., Kimura T., Yamawaki H. and Ikeda K. (1989), "High-T_c BiSrCaCuO superconductor grown by CVD technique," *IEEE Trans. on Magnetics* 25, 2470.

7. Ko J. W., Lee S. Y., Kim H. D., Ha K. H., Ahn J. H., Chung H. S., Togano K. and Maeda H. (1994), "Fabrication of $Bi_2Sr_2CaCu_2O_x$/Ag superconducting tape using a screen-printing method," *J. Mater. Sci.* 29, 4639.

8. Besenyei E., Katona G., Arato P. and Kele A. (1989), "The effect of heat treatment on the superconducting properties of Bi-Sr-Ca-Cu thick films on alumina substrates," *Supercond. Sci. Technol.* 2, 220.

9. Zhou W., Metcalf P. A. and Sato H. (1992), "Bi-Sr-Ca-Cu-O superconducting films prepared by in-situ melting," *J. Mater. Proc. Technol.* 29, 331.

10. Ochsenkuhn-Petropulu M., Tarantilis P., Parissakis G., Psycharis V., Moraitakis E. and Niarchos D. (1994), "Preparation and characterization of electrophoretically deposited high-temperature superconducting coatings," *Microchimica Acta,* 113, 19.

11. Sarkar P., Mathur S., Nicholson P. S. and Stager C. V. (1991), "Fabrication of textured Bi-Sr-Ca-Cu-O thick film by electrophoretic deposition," *J. Appl. Phys.* 69, 1775.

12. Pont M. and Muñoz J.S. (1993), "Superconducting thick films deposited by plasma spray on metallic substrates," in *Superconductivity in Spain.* Félix Ynduráin, ed. Madrid, Spain, 417.

13. Pawlowski L., Hill A., McPherson R., Garvie D., Przelozny Z. and Finlayson T. (1990), "Properties of plasma sprayed $YBa_2Cu_3O_x$ high temperature superconductors," *Proc. 3rd Natl. Thermal Spray Conf. ASM,* Long Beach, California, 641.

14. Sankawa I., Konaka T., Matsuura T. and Ishihara K. (1988), "Superconductivity in $YBa_2Cu_3O_{7-x}$ thick films by the atmospheric plasma powder spraying," *Jap. J. Appl. Phys. Lett.* 27, 1083.

15. Karthikeyan J., Sreekumar K. P., Kurup M. B., Patil D. S., Anantapadmanabhan P. V., Venkatramani N. and Rohadgi V. K. (1988), "Plasma-sprayed superconducting $YBa_2Cu_3O_{7-x}$ coatings," *J. Phys. D: Appl. Phys.* 21, 1246.

16. Mori N., Itoi Y. and Okuyama M. (1989), "Superconducting Y-Ba-Cu-O thick films on silicon and alumina substrates prepared by the plasma spraying method," *Jap. J. Appl. Phys.* 28, 239.

17. Konaka T., Sankawa I., Matsuura T., Higashi T. and Ishihara K. (1988), "Preparation of Y-Ba-Cu-O superconducting tape by atmospheric plasma spraying," *Jap. J. Appl. Phys. Lett.* 27, 1092.

18. Mamalis A. G., Szalay A., Pantelis D. I., Pantazopoulos G., Kotsis I., and Enisz M. (1995), "Excimer surface treatment of high-T_c superconductive ceramic layered plasma-sprayed silver plates," *Supercond. Sci. Technol.* 8, 470.

19. Lungsheider E. and Weber T. (1990), "Plasma-spraying of Bi-Sr-Ca-Cu-O high temperature superconductors," *Proc. 3rd Natl. Thermal Spray Conf. ASM,* Long Beach, California, 635.

20. Bouanani A. M., Suryanarayanan R., Gorochov O. and Brun G. (1989), "Thick superconducting coatings of YBaCuO and BiSrCaCuO," *Proc. Int Conf. Modern Aspects of Superconductivity,* Paris, France, 283.

21. Tönshoff H. K., Butje R., Kñnig W. and Trasser Fr-J. (1988), "Excimer laser in material processing," *Annals of CIRP* 37/2, 681.

22. Kotsis I., Enisz M. and Kristof J. (1989), "Preparation of an $YBa_{2-x}K_xCu_3O_y$-type superconductor with a single firing process," *Proc. 32nd IUPAC Congress,* Stokholm, Sweden, 107.

23. Rosiwal S. M. and Bergmann H. W. (1990), "Surface treatment with the excimer laser and quality control by means of difference reflectometry," *Proc. 3rd Int. Conf. Laser Treatment of Materials,* Erlangen, Germany, 895.

CHAPTER 7

Characterization of HTS Powders and Components

7.1 GENERAL

A wide spectrum of testing methods are used to characterize the high-T_c superconductive materials to optimize the preparation techniques and, therefore, the quality of the products that are used in many engineering applications.

Strength and toughness are the main mechanical properties of most engineering machines or structural elements used to ensure the function of the machine or the stability of the structure, which must be determined. Specific physical and chemical properties that engineering materials should possess depend on the sort of their practical application. Chemical stability, phase purity, microstructural homogeneity and high density are very important physicochemical properties for ceramic superconducting materials. The enhanced electrical and magnetic properties and, therefore, the good function of superconducting components are based on the chemical and microstructural characteristics, which may be changed according to the manufacturing process used.

The main characterization methods for determining chemical (chemical and structural characterization) and physical quantities (physical characterization) dealt with the performance of superconductivity are outlined below.

7.2 CHEMICAL AND STRUCTURAL CHARACTERIZATION

7.2.1 Microscopy and Microanalysis

OPTICAL MICROSCOPY

The optical or light microscope is a powerful tool for examining, evaluating and quantifying the microsructure of materials. The practical maximum magnification of an optical microscope lies between 1000 and 1500×, and it is limited by the wavelength of the visible light.

Careful specimen preparation is necessary for optical metallographic examination. This preparation procedure consists of the following stages:

- Sectioning: This stage is the isolation of a specimen's surface for further examination. This process must be performed slowly, to avoid thermal effects, by using a diamond-cutting wheel or a microtome.
- Mounting: The encapsulation of the specimen into a resin mold of certain dimensions.
- Polishing: This is the main stage of sample preparation, and it is very often performed by mechanical means, known as mechanical polishing. Abrasive SiC grinding paper is used for the initial polishing steps followed by fine polishing using alumina or diamond media in paste or spraying form; oxide (silica) polishing is often suggested as the final polishing step. The initial polishing steps (SiC grinding) are performed in dry conditions or by using butanodiol as lubricating medium. The HTS ceramic is often decomposed by the presence of water. The final polishing steps, e.g., fine polishing or finishing, are performed by using an alcohol-based solution as a lubricating medium. Careful inspection of the specimen under the microscope after every polishing step is necessary.
- Chemical etching: This is a structure-revealing process by immersing the specimen in a chemical reagent, necessary for the metallographic preparation of metals and alloys. However, the etching process is not often met in the HTS sample preparation; dilute halogen anion-alcohol base solutions are sometimes used for chemical etching in multiphase BSCCO compounds.

Various types of microscopical illumination can be used for HTS material characterization. The most prominent imaging technique is that provided by using polarized light microscopy. In multiphase HTS ceramic systems the various crystallographic phases are characterized by differential contrast under polarized light illumination. The results of the morphological mapping can be treated by using image analysis techniques for quantitative evaluation of the microstructure, e.g., phase and porosity percentages, grain size, etc.).

SCANNING ELECTRON MICROSCOPY (SEM)

The scanning electron microscope provides an extensively wide spectrum of magnifications for microstructural evaluation ranging from 50 to often more than 20,000×. The useful magnifications for HTS samples lie practically between 1000 and 5000×.

The electron guns are operated under high-accelerating voltages, varying between 10 and 60 kV. The electron beam is created by filament thermionic emission and is directed and converged by the use of magnetic coils in a vacuum column. In SEM, fractured and polished surfaces can be observed. Conductive coatings, made of gold or graphite, are usually sputtered on the sample surface before examination. Preparation demands are not quite as high as in optical microscopy due to higher depth of field provided by SEM.

Different types of imaging can be realized by SEM observations: secondary electron imaging (SEI) for topographic and morphological evaluation and backscattered electron imaging (BSE) for topographic and chemical (compositional) examination.

TRANSMISSION ELECTRON MICROSCOPY (TEM)

Transmission electron microscopy constitutes one of the most powerful instruments for investigating the micro- and nanostructure of the HTS materials; magnifications over 100,000× can be achieved. A great variety of atomic scale defects, i.e., nanoscale twins, dislocation substructures, stacking faults, grain-boundary inhomogeneities and secondary phase precipitates can be revealed by TEM.

The sample preparation techniques are based on the production of a thin-electron transparent foil of thickness less than 500 nm; the principal techniques are electrolytic thinning and ion bombardement. Further details dealing with the electron microscopy techniques can be found in Reference [1].

FIELD ION MICROSCOPY (FIM)

Field ion microscopy is extensively used for investigating the structure of conducting solids, because [2]:

- The FIM makes possible the study not only of the atomic scale structure of solids but also the surface diffusion of single atoms and their complexes on a smooth surface, as well as the early stages of chemical reactions in metal surfaces.
- Atom probe FIM is the main tool for 3D analysis of the microstructure and chemical composition of a wide spectrum of commercially available materials.

FIM is a unique technique for studying the physicochemical processes taking place on a HTS surface at cryogenic temperatures, under conditions of a high-intensity positive polarity electrical field of the order of 10^8 V/cm. To obtain such high electrical fields, the sample to be examined is made in needle-shaped form with a radius of curvature at the apex of 10–100 nm. The sample is placed in an ultrahigh vacuum chamber cooled down to a cryogenic temperature. An imaging gas, such as nitrogen and argon, is then introduced into the system. When the potential difference between the sample and the fluorescent screen atain a certain level (5 kV), the electric field on the sample's surface becomes sufficient for the filed ionization of the imaging gas atoms. The ion pattern observed on the screen of the microscope consists of bright points, each of them representing single surface atoms at the tip apex; see Figure 7.1.

ENERGY DISPERSIVE X-RAY ANALYSIS (EDAX)

The energy dispersive spectrometry is a semiquantitative analyzing technique for tracing chemical elements in local positions on the sample surface. This technique is applied in parallel to the scanning electron microscopy.

The X-rays emitted by the sample pass into a standard spectrometer through a beryllium window, where they are collected by a solid-state detector maintained at liquid nitrogen temperature. The EDAX spectrum shows the variation of intensity, which is a measure of the X-ray detecting signal, as a function of energy in keV, see Figure 7.2. The energy axis (in keV) corresponds to the elemental electron shell transition energies, usually of K_{α} or K_{β}. X-ray mapping of the chemical elements participating in a local specimen surface can be derived by tracing the fundamental $K_{\alpha,\beta}$ or $L_{\alpha,\beta}$ transitions. The prevention, or even reduction, of certain energies by the window limits the tracing abilities of the analyzer; elements with an atomic number lower than 11 cannot be detected.

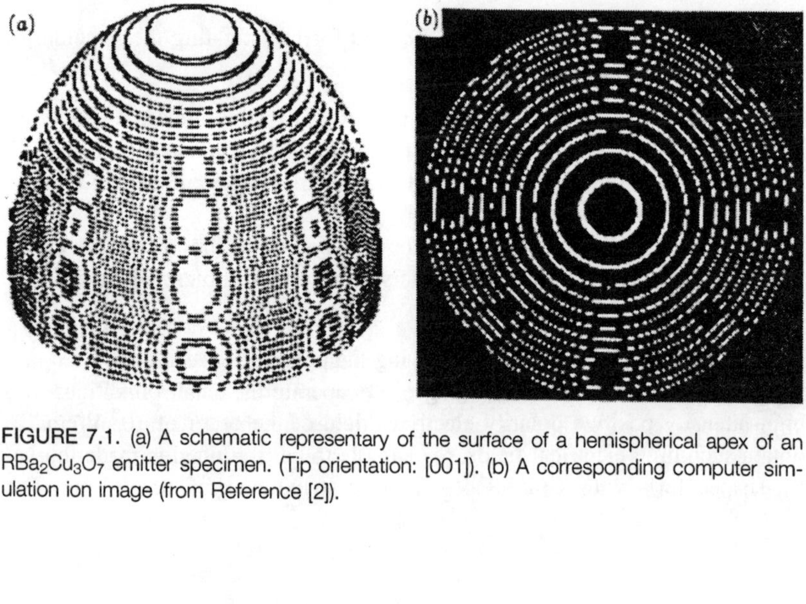

FIGURE 7.1. (a) A schematic representary of the surface of a hemispherical apex of an $RBa_2Cu_3O_7$ emitter specimen. (Tip orientation: [001]). (b) A corresponding computer simulation ion image (from Reference [2]).

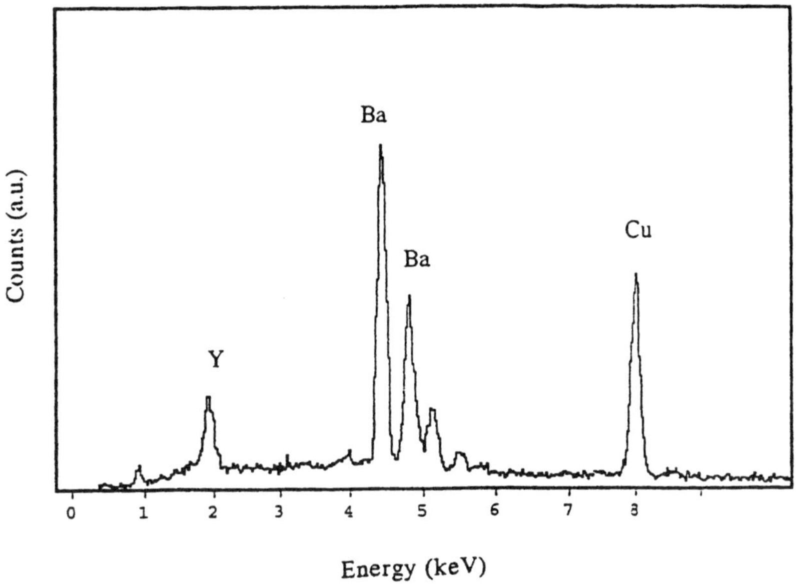

FIGURE 7.2. A typical EDAX spectrum of a Y-Ba-Cu-O superconductor.

7.2.2 Diffraction Techniques

X-RAY DIFFRACTION (XRD)

X-ray diffraction technique is, perhaps, the most important method for evaluating the crystal structures existing in a material. X-rays are photons of wavelength of the order of a nanometer. An incident X-ray beam on a material penetrates into the solid material for many micrometers, and the intensity in the direction of the diffracted beam is determined by the periodicity of the atom planes in the crystalline solid. The consideration of such an X-ray beam of wavelength λ, incident to a crystalline material at an angle θ and diffracted also at the same angle, is illustrated in Figure 7.3. The interplanar spacing of the material is equal to d. The necessary condition for diffraction is known as the *Bragg equation:*

$$2d \cdot \sin\theta = n\lambda,\ n = 0,1,2\ldots \tag{7.1}$$

where the integer n is known as *diffraction order.*

By measuring the angle, θ and knowing the wavelength, λ, the interplanar spacing may be determined and, thus, the crystal structure may be revealed and indentified. To enable identification of most solid compounds, a large li-

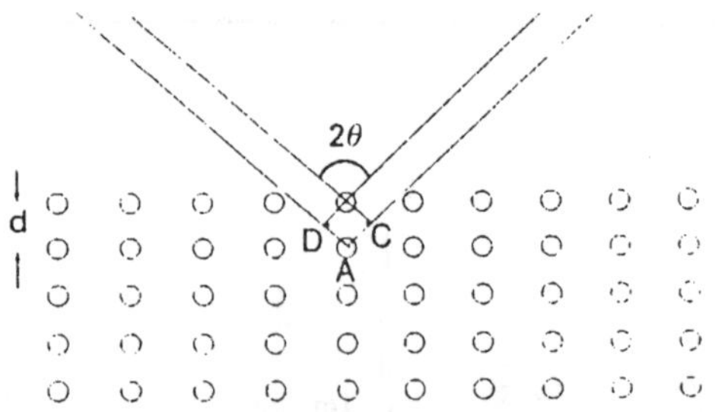

FIGURE 7.3. Principles of X-ray diffraction in a single-crystal material.

brary of expected diffraction patterns has been stored. The X-ray diffraction technique applied to multiphase material is mainly qualitative, but it could also be quantative by appropriate diffraction data processing.

Besides the information about the crystalline phases, the analysis of XRD peaks may provide also useful information about the crystallinity, the crystallographic orientation and the grain size of the material to be examined.

ELECTRON DIFFRACTION

Electron diffraction is an accompanying technique to TEM, exhibiting very high resolution, for investigating the crystal structure of microstructural features revealed by microscopic observations. Theories dealt with the kinematic and dynamic character of electron diffraction will not be discussed here; for details see Reference [1].

Leaving the gun, the electrons are formed by the lens into a crossover, and this demagnified source image is projected onto the specimen by condenser lenses. The first condenser lens forms a demagnified image of the source of diameter about 1 μm that is subsequently projected onto the specimen by a second condenser lens by a magnification of about two. The final illumination electron beam may be as small as 2 μm in diameter, which is sufficient to fill the screen at higher magnifications. For a parallel incident electron beam, the diffracted beams leaving the specimen are focused on the back focal plane of the objective lens. The diffraction pattern is observed, if the back focal plane is projected on the viewing screen by reducing the excitation of the first projector lens. The electron diffraction pattern consists of illuminated spots, which compose the diffraction atom planes of the crystal. The interpretation of electron diffraction patterns is not easy and requires a lot of experience in crystallography.

7.2.3 Spectrometric Techniques

RAMAN SPECTROSCOPY

The *Raman effect* is the interaction between monochromatic light, emitted generally from a laser source, and the chemical bonds within a specimen, which produces a result similar to infrared absorption. When a laser beam (usually Ar or He-Ne) impinges on a material, most of the light is scattered at the same wavelength, known as *Rayleigh scattering,* but a small number of photons may excite molecular vibrations in the specimen. These photons will lose an amount of energy equal to that imparted to the specimen and, therefore, will be scattered with a slightly longer wavelengh. This difference in the wavelength is known as *Stokes shift.* When the excited atoms in the specimen subsequently relax, they release energy back to the incident beam which is scattered with a slightly shorter wavelength, known as *anti-Stokes shift.* Because the change in wavelength of the Stokes and anti-Stokes lines is extremely small, about $\lambda/100$, and, the intensity is considerably reduced compared with the incident laser beam; it is necessary to use very sophisticated equipments to detect these photons. The depth of the material analyzed by Raman spectrometry is of the order of the fraction of laser beam wavelength [1].

An optical microscope device is necessary for parallel observation. Laser light passes through focusing optics and a dichroic beamsplitter to the specimen in the optical microscope. The scattered light then passes through the beamsplitter and filters to a cooled camera detector (CCD). Raman spectra of HTS crystals may provide useful information about atomic bonding, oxygen deficiency and the effect of doping on atomic arrangement [3].

MÖSSBAUER SPECTROSCOPY

This technique has been discovered by Mössbauer in 1958 (Nobel Prize in Physics). *Mössbauer spectroscopy* involves the emission and absorption of γ-rays to determine the chemical state of atoms in a lattice, and it is primarily used to study iron-containing materials. The nucleus of a radioactive isotope is in an unstable condition and decays with the emission of alpha and beta particles, neutrons and gamma rays. If the gamma ray is incident on an absorber of the same element and isotope as the emitter but with the atom in the ground state, then the γ-ray is absorbed and the atom is raised to an excited level. When the γ-ray is emitted, the decaying atom should recoil, whereas a recoil should be also in the absorbing nucleus as it absorbs the γ-ray.

Mössbauer realized that, for the effect to occur, both the emitting and absorbing atoms must be bound in a lattice, where the lowest vibrational excitation energy is greater than the nuclear recoil energy. In such a condition, the atom is unable to recoil, and the γ-ray may be emitted and absorbed without energy loss. This requires that the atom must be fixed in a solid crystal lattice below the Debye temperature. Even when the recoil has been attained, the γ-ray will not be absorbed if the absorbing atom does not possess the same energy gap, between the ground and the excited energy state, as the emitting atom. This situation will occur when the absorbing atom is chemically bound with other elements, which produces small shifts in the energy levels. The Mössbauer effect has limited applications to HTS materials and especially when Fe-doping is used in structure modification.

7.2.4 Thermal Analysis

DIFFERENTIAL THERMAL ANALYSIS (DTA)

Differential thermal analysis is based on the evolution of temperature difference as a function of time between a reference sample and the HTS sample under investigation. The heating rate of both samples is identical. The sample under investigation possesses chemical transformation in the examination temperature range. The ΔT-T plot is known as *thermograph.* When a phase change takes place on heating or cooling, the temperature of the sample deviates from the reference temperature or baseline, and it remains constant until the reaction is completed. Endothermic reactions, e.g., melting, eutectoid transformations, are recorded as valleys in the thermograph, whereas exothermic reactions are recorded as peaks in the same thermograph. The area of the DTA valleys or peaks correspond to the heat due to reactions. Indirect chemical composition measurements and kinetic studies can be performed by means of DTA [4].

DIFFERENTIAL SCANNING CALORIMETRY (DSC)

Differential scanning calorimetry is similar to DTA and it is usually regarded as a more quantitative technique. In DSC the sample and the reference material are maintained isothermal to each other, by controlling the heating and cooling rate at a linear rate via electric energy. The resulting curves of DSC are a recording of heat flow as a function of temperature. As in DTA, the area enclosed in DSC peak is directly proportional to the enthalpy change. DSC can be easier used for quantitative analysis of materials because the calibration coefficient is independent of temperature for DSC.

THERMOGRAVIMETRIC ANALYSIS (TGA)

Modern thermogravimetric analysis instruments consist of a precision thermobalance, a programmable furnace, and a computer to control the equipment and process data. TGA measures the mass of the specimen as a function of the sample temperature. The resulting TGA graph, i.e., the mass change vs. temperature, which is in the shape of stepwise curves, provides useful information about phase indentification and chemical processes, such as oxidation, reduction, decomposition, volitilization, sublimation and environmentally induced reaction. Useful information on oxygen deficieny in YBCO ceramics can be also provided by TGA curves.

7.3 PHYSICAL CHARACTERIZATION

7.3.1 Electrical Characterization

RESISTIVITY MEASUREMENTS

The measurement of resistivity of HTS materials is very similar to the one applied in metals; the main difference is the abrupt elimination of electrical resistance below T_c, whereas in metals the same parameter decreases continuously. The resistivity is determined by measuring the electrical resistance and the sample dimensions. Very small values of resistance are difficult to be measured. Usually, zero resistance could be considered below a certain upper limit. The four-probe technique is principally used for the measurement of resistance as a function of temperature. The two end leads are current leads, whereas the two intermediate leads are used for voltage drop measurement. Current intensities are of the order of mA, whereas voltage drops are of the order of nV. A lock-in amplifier is used for magnifying the recorded signal. Abrupt voltage (or near zero) measurement transition declares zero resistance and, therefore, the onset of superconducting behavior [5].

The attachment of the four contacts, made of copper, silver or gold, for resistance measurements, renders considerable difficulties. These connections must be clean and sound and possess sufficient mechanical strength, to avoid the addition of errors originated from metal lead-ceramic bonding. The metal connections are usually made by soldering using a silver-based paint. To achieve lower connection impedance, contact material is sputter deposited on the clean HTS surface and subjected to a subsequent oxygen annealing at 600°C [6].

CRITICAL CURRENT DENSITY (J_C) MEASUREMENTS

The same four-probe technique is applied for critical current density, J_c direct measurements. A powerful current supplier is necessary for these measurements. The *V-I* curve for a HTS sample of known dimensions is taken at a certain temperature T ($T < T_c$) and applied magnetic field ($H < H_{c2}$). Below a critical current intensity the sample behaves as a superconductor; i.e., the voltage drop is nearly zero. The zero voltage is considered when the measured value is lower than a critical value, usually of 1 μV/cm (known as *1 μV/cm criterion*). Above the critical current intensity value the sample is no more superconductive exhibiting ohmic behaviour.

7.3.2 Magnetic Characterization

DC-MAGNETIZATION MEASUREMENTS

The magnetization and magnetic susceptibility measurements are based on the diamagnetic behavior of HTS materials. The characterization leads to the estimation of T_c and ΔT_c. However, sophisticated models, such as flux-pinning and glassy-state, can be implied to explain the superconducting behavior using remanent magnetization measurements [7].

The most commonly dc-magnetometers, such as VSM (vibrating sample magnetometer) and SQUID, measure the magnetic moment of the sample. If the sample to be measured does not have a permanent magnetic moment, an applied field is required to magnetize it. A detection coil is then used to detect the change in magnetic flux, due to the presence of the magnetic moment. The sample flux, coupled to the detection coil, is made to vary by the sample movement. In a VSM system, the sample is vibrated near the detection coil; as the sample vibrates, an ac-signal is generated at a frequency determined by the sample oscillation. In a SQUID system, the sample only passes through the detection coil. The dc-magnetic susceptibility, χ_{dc} is given by the equation

$$\chi_{dc} = \frac{M}{H_{dc}} \tag{7.2}$$

The magnetization M is given as a function of magnetic moment, *m,* as

$$M = \frac{m}{V} \tag{7.3}$$

where V is the volume of the sample.

AC-MAGNETIC SUSCEPTIBILITY MEASUREMENTS

As in the dc-magnetometers an ac-susceptometer uses the detection coil to detect changes in the magnetic flux due to the sample. In an ac-susceptometer the sample is generally entered within a detection coil and driven with an applied ac-magnetic field, see Figure 7.4. Movement of the sample is not required to produce an output signal. The actual ac-susceptibility, χ_{ac} is a two component vector

$$\chi_{ac} = \chi' + \chi''j \tag{7.4}$$

The real component, χ', represents the component of the susceptibility, which is in-phase with the applied ac-field, whereas the imaginary component, χ'' represents the out-of-phase part of the ac-susceptibility. The real part is often called *diamagnetic susceptibility* and is equal to the derivative dM/dH, whereas the imaginary part is related to the energy or hysteresis losses due to the magnetic field cycling. Information for the quality of the superconducting material, apart from T_c and ΔT_c, can be provided by ac-susceptibility measurements, regarding the inter- and intragranular coupling, anisotropy, frequency dependence behavior, relaxation processes and experimental verification of the critical state models; for details see References [9] and [10].

MAGNETO-OPTIC IMAGING

Magneto-optic imaging is a relatively new technique by which the magnetic flux distribution patterns in a superconducting material can be obtained [10,

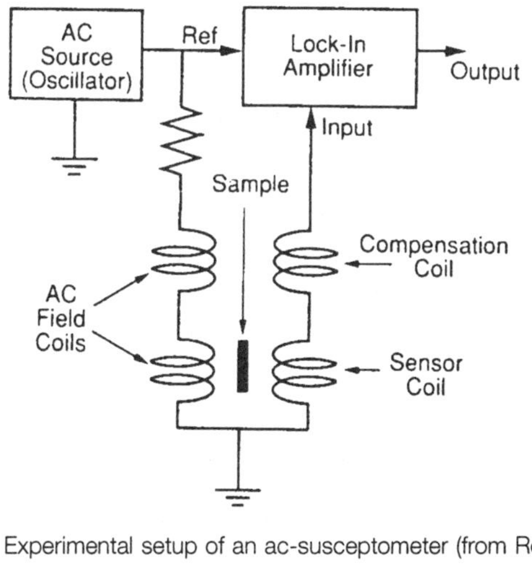

FIGURE 7.4. Experimental setup of an ac-susceptometer (from Reference [5]).

11]. In an appropriate magneto-imaging apparatus real-time observations of the dynamics of spatial distribution in a superconductor can be performed. The Faraday effect in a thin crystalline film is used to visualize the flux distribution in a superconductor, placed in close contact to the indicator film (usually a doped iron-garnet crystal). Where magnetic flux is present in the superconductor, light is reflected, after passing through the indicator material twice by an angle proportional to the flux density. By using a microscope with crossed polarizers, an image is seen, where the contrast directly represents the flux distribution; dark areas contain little flux, and bright areas are dense.

Recent developments of the magneto-optic visualization technique have proved to be extremely useful in the basic studies of flux penetration and pinning in HTS materials. Moreover, the technique is highly effective to localize material defects, such as microcracks, grain boundaries and secondary phases, which may have undesirable effects on the functional performance of superconducting components used in engineering applications.

7.4 REFERENCES

1. Flewitt P. E. J. and Wild R. K. (1994), *Physical Methods for Materials Characterisation,* IOP Publishing Ltd, London.
2. Talantsef E. F. (1995), "Field ion microscopy of high-temperature superconductors," *Supercond. Sci. Technol.* 8, 593.
3. Pantazopoulos G. (1997), "Fabrication of metal-sheathed ceramic superconductors by explosive compaction/cladding and subsequent forming," PhD thesis, Dept. Mechanical Engineering, National Technical University of Athens, Greece (in Greek).
4. Geiger G. (1993), "Thermal for improved quality," *Am. Ceram. Soc. Bull.* 79, 54.
5. Doss J. D. (1989), *Engineer's Guide to High-Temperature Superconductivity,* John Wiley & Sons, New York.
6. Ekin J. W., Larson T. M. and Bergren N. F. (1988), "High-T_c superconductor/noble metal contacts with surface resistivities in the 10^{-10} Ω cm^2 range," *Appl. Phys. Lett.* 52, 1819.
7. Malozemoff A. P., Krusin-Elbaum L., Cronemeyer D. C., Yeshurun Y. and Holtzberg F. (1988), "Remanent moment of high-temperature superconductors: implications of flux-pinning and glassy models," *Phys. Rev.* B38, 6490.
8. Mei Y., Luo H. L., Chen D. X., Nogues J. and Rao K. V. (1988), "AC magnetic susceptibility of $YBa_2Cu_3O_{7-x}$," *J. Appl. Phys.* 64, 2533.
9. Zakharchenko I. V., Nicolaides G. K., Rao K. V. and Muhammed M. (1991), "Grain orientation and anisotropic superconducting properties of textured bulk $YBa_2Cu_3O_x$," *Physica C* 185–189, 2295.
10. Koblischka M. R., Johansen T. H. and Bratsberg H. (1997), "Bending of silver-sheathed (Bi,Pb)-2223 tapes investigated by magneto-optical flux visualization," *Supercond. Sci. Technol.* 10, 693.
11. Koblischka M. R., Johansen, T. H., Bratsberg, H. and Vase P. (1998), "Study of flux entry and exit into Bi-2223 multifilamentary tapes," *Supercond. Sci. Technol.* 11, 479.

CHAPTER 8

Industrial Applications

8.1 GENERAL

Superconducting materials possess electrical/electronic and magnetic properties of utmost importance, one of which is the high current-carrying capability of superconductors with small cross sections. For low-temperature superconductors, mainly Nb-Ti and Nb_3Sn, a refrigeration process in liquid helium is necessary; this refrigeration system is rather complex using also a very costly coolant like helium. The new high-temperature ceramic superconductors exhibit their useful technical properties using a simpler liquid nitrogen refrigeration system.

A wide range of electrotechnical applications using HTS components, from HTS computer chips to high-power transmission cables and levitated vehicles, may be listed. In the present book such HTS applications are divided into two main groups:

- Bulk applications: In this case bulk ceramic superconducting materials are used as long power transmission cables or other functional elements in electrical machines, such as conductors and transformers, see Figure 8.1.
- Film applications: In this case HTS thin and thick films are used as elements in microwave cavities, IR detectors, SQUIDS and memory devices for high-speed computers, see Figure 8.2.

The basic features of the superconducting and the design followed for their construction will be analyzed further. An overview of the engineering applications of HTS is presented in Reference [1].

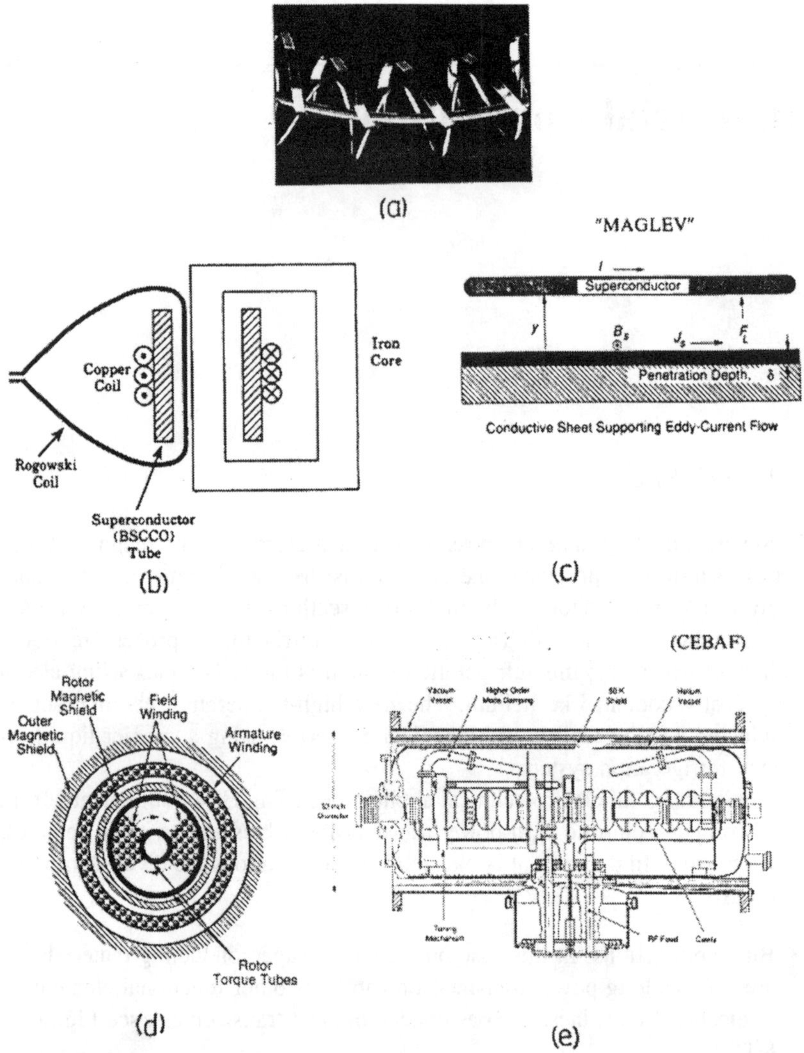

FIGURE 8.1. Applications of bulk superconducting materials. (a) wires, coils, etc.; (b) magnetic shielding (fault-current-limiter); (c) magnetic levitation (MAGLEV); (d) HTS generator; (e) superconducting accelerator.

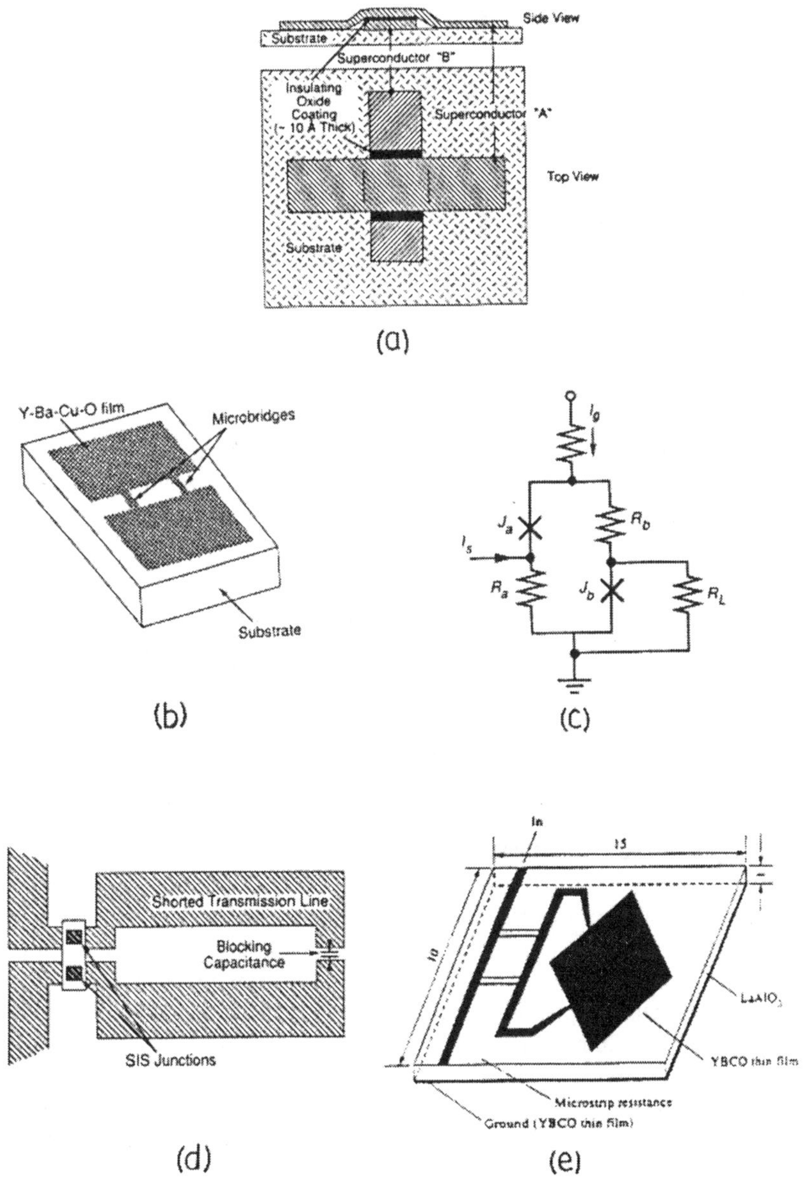

FIGURE 8.2. Applications of superconducting films. (a) Josephson junction; (b) SQUID; (c) logic gate; (d) S-I-S mixer; (e) HTS microwave antenna.

8.2 SMALL SCALE ELECTROMAGNETIC HTS MACHINES

8.2.1 Design Principles

When designing an experimental machine the following may be considered, see References [2] and [3]:

- The ceramic superconducting materials are brittle and rigid and, consequently, the number of the conventional coil forms is limited. Therefore, HTS forms using an external metal case are used for the construction of electrical machines.
- The design of the machine and the practical solution of the cooling of the HTS parts must be as simple as possible.
- The connections between the superconducting conductors and the other parts of the electrical circuit must possess low contact resistance, and, therefore, HTS ceramics with good electrical contacts are needed.
- Because the current-carrying capability of some types of bulk-processed HTS, such as hot-pressed or explosively compacted components, is relatively low, excitation coils made of HTS materials are not used. Note that for conventional superconducting synchronous generators, the excitation coils are fabricated from conventional metal alloys or intermetalic compounds possessing low-temperature superconductivity.

The basic aspects of three types of electromagnetic machines, namely a synchronous generator, a levitated bearing and a levitated vehicle, based on different fabrication methodologies are outlined below, see also Reference [4].

8.2.2 Synchronous Generator

The construction of this HTSC model electrical machine has been planned in two stages, see References [3]–[7]. During the first stage an aluminium conductor substitutes for the future superconducting coil. However, the design of the machine makes it possible to insert the superconducting conductors into the places of the aluminium conductors, which is performed during the second stage of the work.

A top and a cross-sectional view of the model machine are shown in Figure 8.3. The two main parts of the model are the stator and the rotor made of steel and aluminium, respectively. Because the current-carrying capability of the ceramic sperconductors is relatively low and an air-gap flux density of the order of 0.5–1.0 T is needed, rare earth permanent magnets, e.g., $SmCo_5$, are used. These magnets, with dimensions $26 \times 8 \times 4$ mm, are placed into the slots of the rotor wall, in three rows along two perpendicular diametral planes; all together 4×3 pieces are used. The magnets possess a remanent flux den-

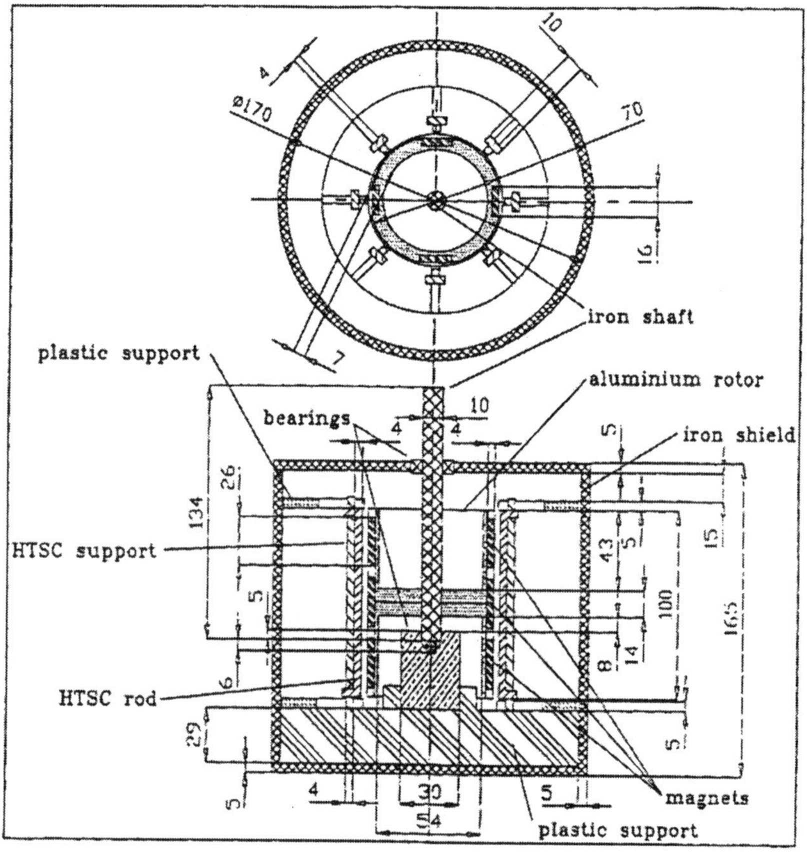

FIGURE 8.3. Schematic top and cross-sectional views of small-scale model synchronous generator (from Reference [4]).

sity of $B_r = 0.9$ T. The 5-mm width of the stator yoke was chosen, so that its magnetic circuit should not saturate. The armature winding, consisting of eight rods, is placed into the gap; from the point of view of the magnetic circuit, the gap must be as small as possible, with its minimum value being determined by the geometry of the HTSC conductor. Note that this winding is made of HTSC conductor of the YBCO or BSCCO compounds, during the second phase of the experimental work.

The working position of the machine is vertical. The shaft of the rotor is placed in a bronze sliding bearing. At the first stage, the cooling of the machine is performed by placing the whole machine into an open liquid nitrogen bath. In the second stage the machine will be cooled in a cryocooler of capacity of 16 W at 70 K.

8.2.3 Levitated Bearing

The schematic cross-sectional view of the HTS magnetic bearing is shown in Figure 8.4; see also Reference [8]. The cylindrical Nd-Fe-B magnets are fitted into the aluminium tube, which constitutes the rotating part of the bearing. Because of the repulsive forces between the magnets and the superconductors, the aluminium tube, together with the magnets, levitates above the superconductors. The trough-shaped superconductors form a "bed" for the rotor. The shaft is driven by an asynchronous motor. The two-pole split-phase stator of the motor is fed from a 220-V, 50-Hz single phase electrical network. Note also that the aluminium tube serves as the rotor of the asynchronous motor. A maximum speed of the shaft up to 900 rpm was obtained.

Three different kinds of HTSC samples in disk form were prepared, namely by cold-pressing BSCCO powder, explosively compacting YBCO powder and melt-texturing YBCO powder.

- Cold-pressed BSSCO disk (sample A): The powder used was made by the traditional solid-state chemical reaction technique, including calcination and heat treatment, using an appropriate stoichiometric mixture of Bi_2O_3, $SrCO_3$, $CaCO_3$ and PbO to give the nominal composition of

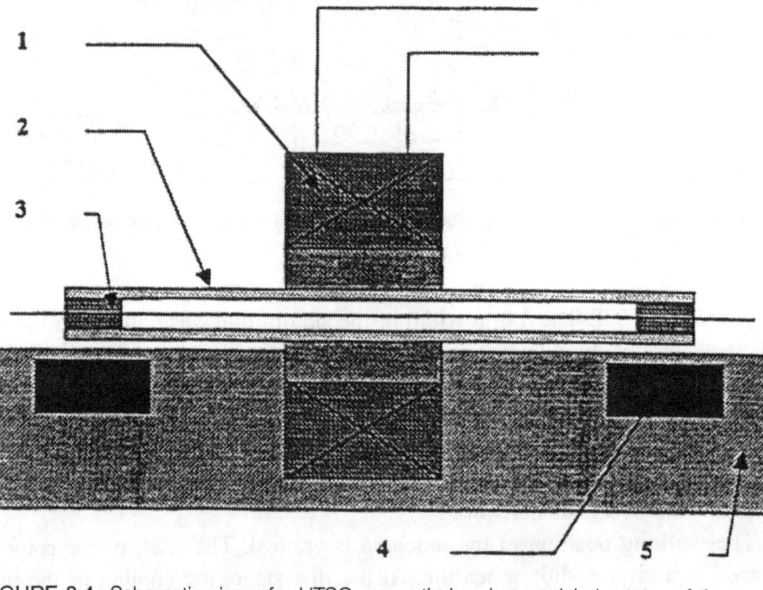

FIGURE 8.4. Schematic view of a HTSC magnetic bearing model. 1, stator of the asynchronous motor; 2, rotating aluminium shaft; 3, cylindrical Nd-Fe-B magnets (rotating part of the bearing); 4, HTSC "bed" (stator part of the bearing); 5, liquid nitrogen (from Reference [4]).

$Bi_{1.5}Pb_{0.5}Sr_2Ca_2Cu_3O_x$. Lead was added to improve the reaction kinetics of the formation of the 2223 phase. Disks with various dimensions, according to the relevant application, were fabricated by using a hydraulic press by applying a pressure up to 20 MPa. The critical temperature, T_c, of the sample was 112 K. More details about the fabrication of Bi-based superconductors are presented in Reference [9], but see also Chapter 5 above.

- Explosively compacted YBCO disk (sample B): The superconducting powder was prepared by the conventional solid-state reaction technique, including calcination and heat treatment, from stoichiometric quantities of Y_2O_3, $Ba(OH)_2 \cdot 8H_2O$, $KF \cdot 2H_2O$, CuO and $AgNO_3$. The final nominal composition of the sample was $YBa_{2-x}K_xCu_3O_v$ (x = 0.01–0.1). Potassium-doped YBCO materials possess improved sinterability, reduced oxygen sensitivity and better physical properties (T_c, J_c); see References [10] and [11] and Chapter 3 above. Silver addition, although Ag is not incorporated in the superconducting (123) compound, helps the oxygenation process and improves the interparticle bonding and, therefore, the mechanical properties of the ceramic, mainly its toughness and thermal fatigue strength [12]. HTSC billets were fabricated by using the axisymmetric explosive compaction technique; the corresponding experimental setup is the same as that used for cylindrical powder compaction experiments [11], described in detail in Chapter 4, see also Reference [13]. Disks of appropriate dimensions were sliced from the original billets. The critical temperature of the sample was 90 K.
- Melt-textured YBCO disk (sample C): A mixture of $BaCO_3$ and CuO was melted at the temperature of 1080°C. The melted mixture was pulverized and mixed with Y_2O_3 and 0.1% wt Pt powder. The addition of small amounts of Pt (<0.5%) alter the interfacial energy of 211-liquid interface, resulting in supression of the coarsening rate and, therefore, the 211 particle size; see Reference [14] and Chapter 5 above. The 211 phase, namely the Y_2BaCuO_5, was finely distributed in the final $YBa_2Cu_3O_7$ matrix in the form of fine submicron inclusions, creating "domain" structure. This technique was used to form pinning centers for the improvement of the levitating capability of the HTS sample. The mixture was pressed hydrostatically under a pressure of 2 GPa, followed by a special heat treatment. The critical temperature of the sample was 94 K.

8.2.4 Levitated Vehicle

The track for the vehicle is built up of $SmCo_5$ permanent magnets of dimensions 4 × 8 × 26 mm, whereas the HTS, with dimensions of 3 × 18 × 20 mm, is fitted into the bottom of the liquid nitrogen container; see Reference [15] for details. The repulsive force, which is developed between the permanent magnets of the track and the HTS material due to the Meissner effect, leads to the expected levitation phenomena.

8.2.5 Modeling of the Magnetic Properties

DETERMINATION OF THE MAGNETIC FIELD

The HTSC conductor designed for the synchronous generator model machine is a rod with a 4-mm diameter and 100-mm length. Because of its size, the magnetic field distribution inside and outside the superconductor is of particular interest. The calculation of the magnetic field and the magnetic forces between the HTSC conductor and the magnets are performed by the computer code "WinTer," based on the finite difference method, see Reference [8.16]. Note that the "WinTer" is capable of modeling the superconductor by using various types of magnetization curves that, for simplicity purposes, are described by analytical functions.

The superconductor model involves also the hysteresis of the material, taking into account its magnetic prehistory. The code is capable of simulating the movement of either the permanent magnets or the superconductors. In this case the subsequent "snapshots" correspond to the subsequent stationary working points, and, therefore, the movement, whereas, at the same time, the magnetic prehistory of an object is simulated as a series of stationary states. Figure 8.5 shows three consecutive "snapshots" taken, when a superconducting rod with a J_c = 6,000 A/cm^2 rotates in the field of a permanent magnet; this calculation simulates the rotation of the rotor of the experimental machine [16].

DETERMINATION OF THE MAGNETIC LEVITATION

The measurement of the levitation forces are conducted by using the experimental setup described in detail in Reference [8]. Calculation of the levitation characteristics, for the levitated bearing and the levitated vehicle, was

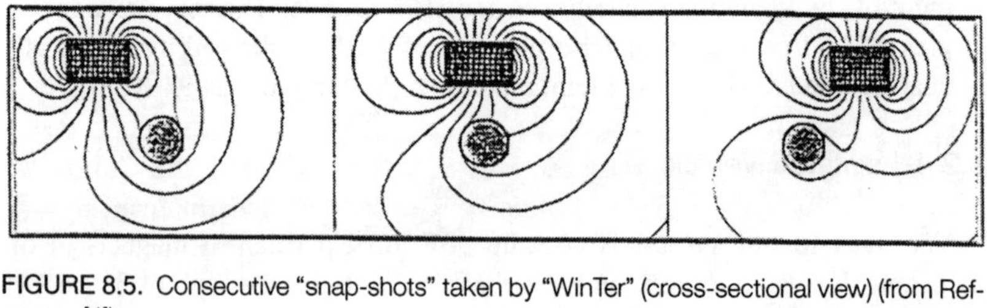

FIGURE 8.5. Consecutive "snap-shots" taken by "WinTer" (cross-sectional view) (from Reference [4]).

accomplished by using the superconductor models and the computer program "WinTer." The input data of the program are the magnetization curve of the various types of the HTSC materials, assuming that, up to the lower critical field, B_{c1}, the HTSC possesses ideal diamagnetic properties and between the lower, B_{c1} and upper critical field, B_{c2}, the HTSC is magnetized uniformly and the volume fraction of the superconducting phase is 100%.

Because the latter condition can be fulfilled only partially, the only input data for this method are the lower critical field, B_{c1}, the field B_D and the "shape" of the curve, i.e., the classification of the HTSC material. B_D is defined as the magnetic flux density related to the maximum value of the magnetization [17]. The classification of the HTSC material, in the frame of this program, was realized according to the analytical "critical-state model" fitting curves, e.g., Bean's or Kim's models. The magnetic field is then calculated as the resultant of the external magnetic field of the superconductor, divided into small elementary volumes influenced by the changing magnetic states of all other elementary superconducting volumes. Calculations are based on the corresponding Maxwell equations for the magnetostatic field [18]. In Figure 8.6

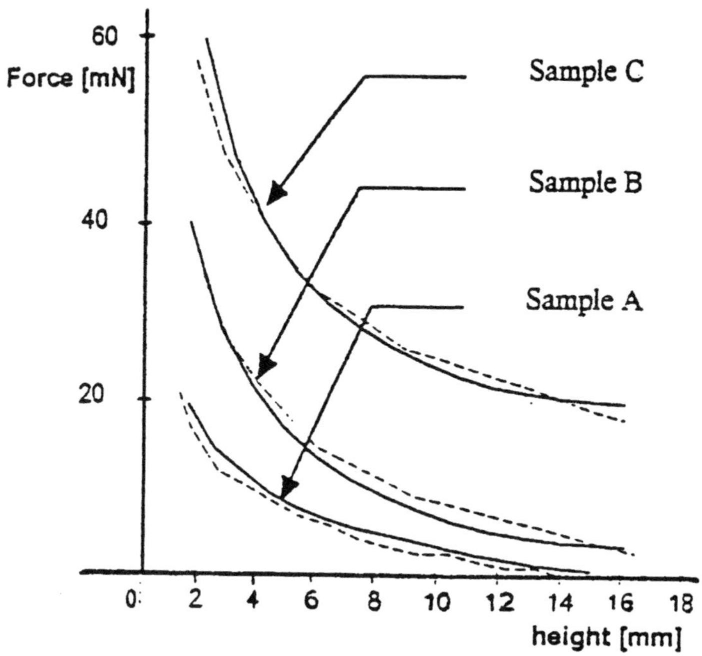

FIGURE 8.6. Experimental and calculated levitating force vs. height curves for various HTSC materials. (solid line, calculated; dashed line, experimental) (from Reference [4]).

experimental and numerical levitation force-height curves for the HTS materials of the BSCCO and YBCO systems, fabricated by the three techniques mentioned above, are shown; very good agreement between the experimental and calculated curves was obtained; see also References [15] and [16].

8.3 LARGE SCALE BULK HTS APPLICATIONS

8.3.1 Superconductive Wires and Coils

FABRICATION OF WIRES

BSCCO-OPIT multifilamentary short-length (about 10 cm) rolled wires at American Superconductor, reached, recently, a core critical current density, J_c of 70,500 A/cm^2, at 77 K, self-field, 1 μV/cm. It is a low, about 10%, HTS fill-factor wire, with 19 filaments in an extra thick silver sheath, designed to achieve maximum uniformity in the HTS filament bundle; note that it is a process that cannot be directly applied to higher-fill-factor long-length manufacturing, see Reference [19] for details. Note also that a 19-filament, intermediate, e.g., 31%, fill-factor wire with a strand or engineering critical current density, J_c of 22,800 A/cm^2 at 77 K, self-field, 1 μV/cm, constitutes an improved commercial approach; the corresponding J_c was 73,000 A/cm^2, at the same conditions. This wire strand has relatively small external cross section dimensions of 0.08 $\times$ 1.78 mm. Critical current densities for BSCCO-OPIT wire, when translated to a practical long-length process, would fulfill most of the expected performance requirements for HTS applications.

During the last several years, there has been increasing attention to an alternative HTS wite technology, namely YBCO-coated conductors, in which biaxial texture is created in a substrate and a YBCO thick film is grown epitaxially on top. The texture results from either an asymmetric thin-film deposition process such as ion beam-assisted deposition or inclined substrate deposition or by deformation texturing of an entire substrate; see Reference [19].

COIL FABRICATION

The coil is fabricated in the form of a saddle coil, e.g., as a single pancake winding with two tapes in parallel, with electrical insulation into the coil structure for layer to layer voltage protection. Voltage taps were installed in eight locations, four around the inner most turn and four around the outer most turn, see Figure 8.7. The winding procedure was based on existing winding technology used for Nb-Ti quadrupole magnet fabrication, see Reference [20], whereas the tooling to support the winding mandrel was designed and fabri-

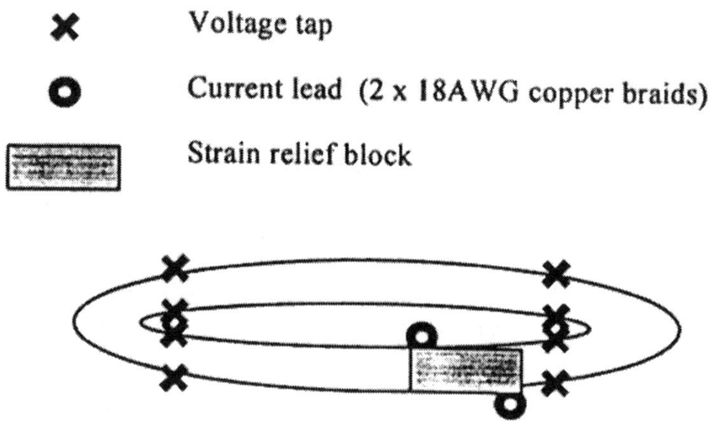

FIGURE 8.7. Voltage tap configuration for comparing coil sections in coil fabrication (from Reference [20]).

cated by the American Superconductor Corporation. As the tapes were laid around the end turns, a mandrel was pivoted simultaneously to maintain perpedicularity of the winding surface along the eight straight elements. The performance of the coil at 77 K is shown in Figure 8.8, reproduced from Reference [20].

8.3.2 Superconducting Magnets

The construction of superconducting magnets in solid rectangular form or coil form supports many conventional and nonconventional applications, ranging from nuclear reactors to aerospace applications. These specific applications are based on the high magnetic field induced by the supercurrent flow. The current flow in the HTS magnet must be kept below the critical current value. As long as the upper critical field, H_{c2} is not exceeded and the cryogen supply is sufficient to hold the temperature below the T_c, corresponding to the external magnetic field, the operation seems to be easily performed. Note, however, that the manufacturing of good-quality superconductors, with low porosity and close to 100% of superconducting phase, is very critical for high magnetic field applications.

8.3.3 Magnetic Resonance Imaging (MRI)

Magnetic resonance imaging (MRI) is a technique extensively used in medicine, for noninvasive imaging of soft tissues without need for exposing them to ionizing irradiation. The application of superconductivity in MRI is partic-

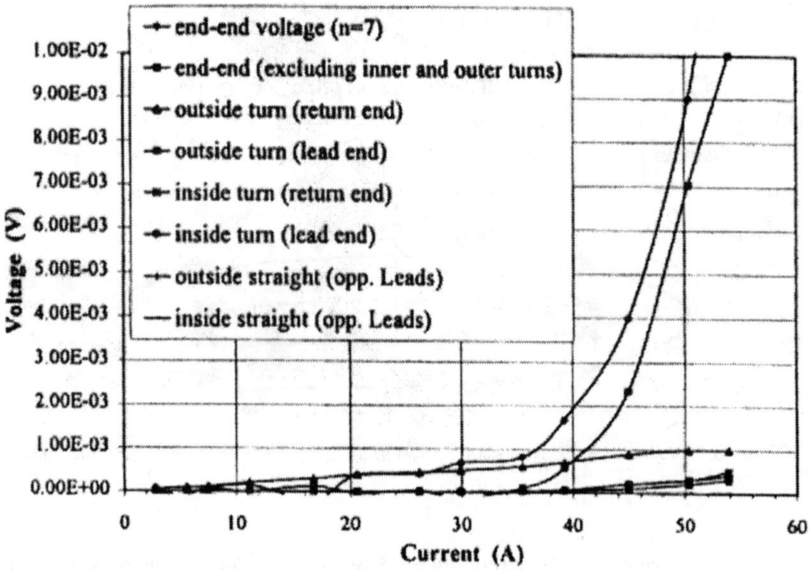

FIGURE 8.8. 77 K performance of the HTS saddle coil (from Reference [20]).

ularly significant, because MRI is the principal market of superconductor magnets [1]. Note that MRI magnets must require very uniform magnetic fields over relatively large volumes, ranging between 10 and 20 Tesla, with field uniformity of about 10 ppm over 0.5–1.0 m distances; such high and uniform fields can be deduced only by superconducting magnets.

The MRI process, known also as nuclear magnetic resonance (NMR), depends on the existence of nuclear magnetic moments, associated with quantized nuclear spins. When a magnetic field is applied, these moments give rise to the distinct Zeeman nuclear energy levels. Applied radiofrequency electromagnetic fields can induce measurable spectroscopic transitions between these nuclear energy levels, leading to the identification of specific nuclei, see Reference [21].

8.3.4 Magnetic Separation

Magnetic separation is a process used for decades in mining and chemical industry. Applications of this process may be found in the elaboration of iron ores and also in the processing of kaolin clay.

This separation process is based on the different magnetic characteristics of the materials to be isolated. If the material to be separated is ferromagnetic, as Fe, Ni or Co (such minerals are magnetite, ilmenite, etc.), relatively weak magnetic fields may be used and there is no need for superconducting magnets. On the contrary, weak magnetic minerals, such as hematite, garnet or limonite, require the use of intense magnetic fields for the separation from a nonmagnetic matrix. The powders to be separated, with particle size ranging from micrometers to millimeters, must be as dry as possible, to enhance the effectiveness of the separation process.

8.3.5 Magnetic Levitation (MAGLEV)

Magnetic levitation possessed by superconducting materials is based on the Meissner effect. Magnetic levitation or suspension of trains of speed over 500 km/h appears attractive because it will minimize the time of the transportation.

There are a variety of ways that magnetic levitation of trains can be achieved. One generic approach uses permanent magnets placed along the track that repel other permanent magnets on the moving vehicle. In a second configuration, electromagnets are used on the vehicle to generate an attractive force via a continuous rail of laminated iron, that is above the onboard electromagnet; this attraction lifts the vehicle above the railbed. A third technique uses onboard superconducting rectangular loops on a continuous conductive sheet on the railbed. As the vehicle moves, eddy currents are generated in the conductive sheet and the resultant repulsion lifts the vehicle, see Figure 8.1(c) and Reference [22]. Taking into account the Laplace law in electromagnetics, the lift force is proportional to the square of the current intensity, flowing the superconducting loop.

8.3.6 Generators

The conceptual design of a HTS generator for mobile radar (MR) applications is described in Reference [23]. The power system for MR consists of a diesel engine, a generator, and power-conditioning electronics. Power density and the overall efficiency of the system are significantly improved if the generator and the power electronics are cryogenically cooled.

The HTS generator, shown in Figure 8.9, is driven by a diesel engine and supplies power from two separate stator windings; the main winding supplies 850 kW at 50 Vrms/phase, and the auxiliary winding 150 kW at 120 Vrms/phase. The central part of the rotor consists of three main concentric structures, the innermost of these sections is the four-field coil steel vessel, with the helium gas coolant, see Figure 8.10, a thermal shield is in the middle concentric component and the vacuum can is the outer concentric component of the rotor. The vessel walls contain insulated vacuum feedthroughs for the field current and instrumentation and connections for helium gas [23].

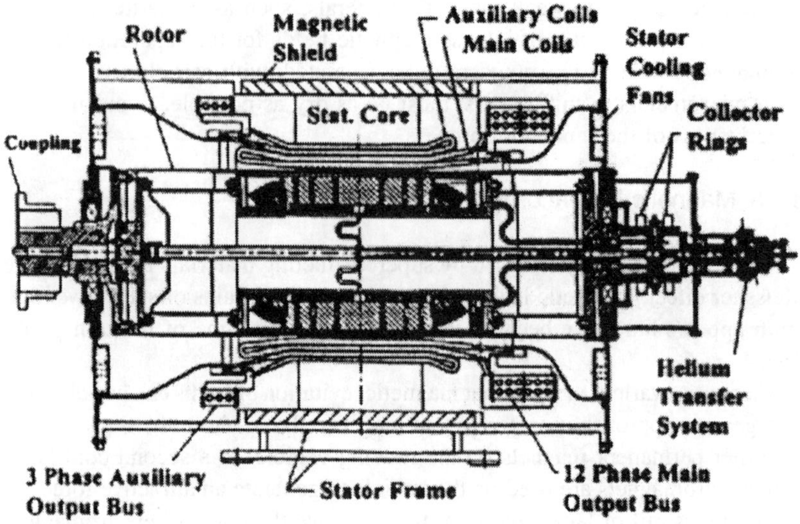

FIGURE 8.9. A cross section of a HTS generator indicating the rotor and stator assembly (from Reference [23]).

8.3.7 Superconducting Magnetic Energy Storage (SMES)

A variety of energy storage techniques have been considered. It may well be that the largest impact of superconducting technology during the beginning of the 21st century will be in superconducting magnetic energy storage (SMES) applications. In the SMES concept the energy is stored in the magnetic field of a rather large superconducting solenoid, see Figure 8.11. For a persistent current I in the SMES inductor L, the stored energy is $LI^2/2$. The solenoid is reversibly connected to the external power system. When the demand is low, the ac line is rectified to supply dc current to the solenoid. During peak demand periods, the dc current in the solenoid is inverted to 60 Hz and fed back into the power grid. An almost 95% efficiency is expected for the SMES systems (most of the 5% loss, about 3%) is expected to be dissipated during the rectification of ac to dc and the inversion back to ac, the balance being primarily due to the energy used in refrigeration of the cryogen. SMES units are also very rapid in response, being able to switch from the storage to the discharge mode in a time on the order of 1 s; such fast response could, in some cases, eliminate blackouts.

8.3.8 Particle Accelerators

Superconductivity technology is applied to particle accelerators in two primary areas, i.e., to save energy and to produce high magnetic fields; there-

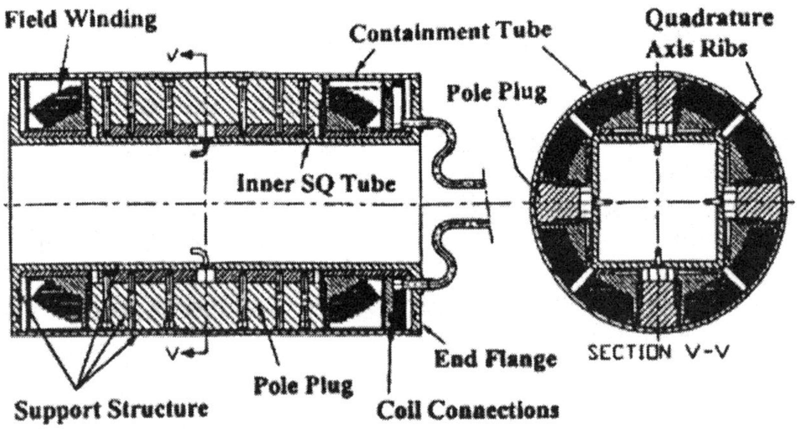

FIGURE 8.10. A cross section of a field coil vessel (from Reference [23]).

fore, the use of superconducting magnets is necessary to use superconductors in the construction of high-quality superconducting accelerator cavities; the primary purpose in the latter case is to achieve high accelerating field gradients and to reduce the radiofrequency power requirements. The proposed superconducting super collider (SSC) would use a large number of superconducting electromagnets. The CEBAF facility (Continuous Electron Beam Accelerator Facility) is also an example of the use of superconductors, see Figure 8.1(e).

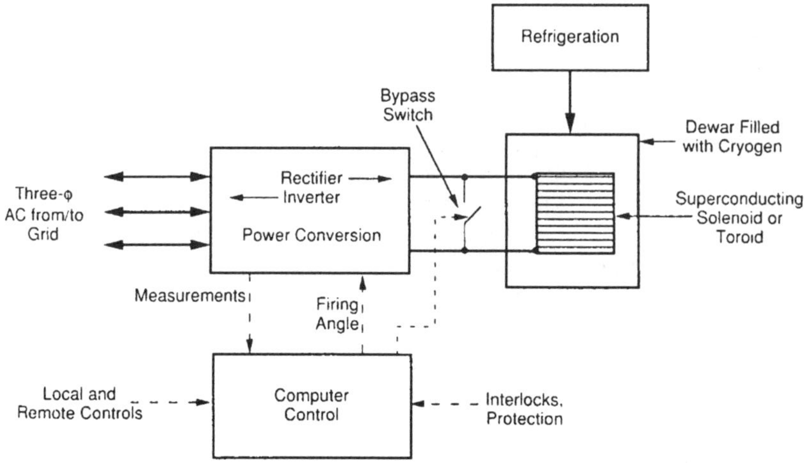

FIGURE 8.11. A schematic representation of the SMES system (from Reference [1]).

Accelerating structures are commonly listed as "high β" or "low β," depending on whether low-mass particles (electrons, positrons) or heavier particles (protons, oxygen ions) are to be accelerated.

8.3.9 Aerospace Applications

Using high-T_c superconductors for space equipment and systems becomes very attractive. A large contactless magnetic support system for applying artificial gravity space station, which provides artificial gravity and zero gravity environments simultaneously, has been proposed. This space station is composed of nonrotating microgravity section at the center, a rotating artificial gravity section and a transit cabin used for transportation between them. The rotating section generates the artificial gravity, whose value depends on the rotating speed and radius. Gravity of 1 g is generated by 3 rpm rotating speed with a 100-m radius. The nonrotating despun section maintains the microgravity environment. The magnetic support system is the main structure in this section, which separates the sections mechanically for vibration isolation and it also acts as the main actuator to control the coning motion produced by gyro effects from the rotating section. The transit cabin is designed to transport the crew and instruments from one section to the other without EVA (extra vehicular activity). Figure 8.12 shows the configuration and a schematic cutoff view of this space station; for more details see Reference [24].

8.4 HTS FILM APPLICATIONS

8.4.1 Josephson Junctions

A Josephson junction is constructed by placing a very thin insulating or normal layer, about 1 nm, at the interface between two superconductors, see Figure 8.2(a). This junction works as an electron pair tunneling device.

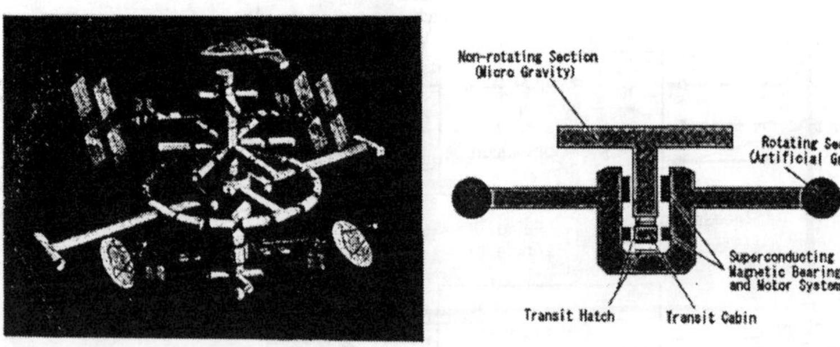

(a) (b)

FIGURE 8.12. (a) Artificial gravity space station; (b) Schematic cutoff view of (a) indicating the superconducting magnetic bearings (from Reference [24]).

Current flow should be below a critical value, I_c, at zero gap voltage. If the current intensity is greater than I_c, then a voltage difference is created at the interface and normal electric current starts to flow. When the critical voltage value is exceeded, e.g., $V > V_g$, the current flow increases rapidly; the Giaever normal electron tunneling is attained, see Figure 8.13 and Reference [1]. The interface current intensity, I ($<I_c$) depends on the phase difference φ ($=\varphi_1-\varphi_2$) between the two sides of superconducting layers. The phases φ_1, φ_2 characterize the Ginzburg-Landau macroscopic wave function of the these two superconducting layers:

$$I = I_c \sin \varphi \tag{8.1}$$

The interface current has an oscillatory form flowing at very high frequencies. For a constant junction voltage, V, the frequency of the current flow is given by the following equation:

$$f = \frac{2eV}{h} \tag{8.2}$$

where, e is the electron charge and h is the Planck's constant.

Josephson junctions are used to a great variety of applications, ranging from very fast switching and memory elements for digital superonductor computer circuitry to sensitive magnetometers such as SQUIDs. Note that in computer applications the Josephson junctions are used for switches, by moving the

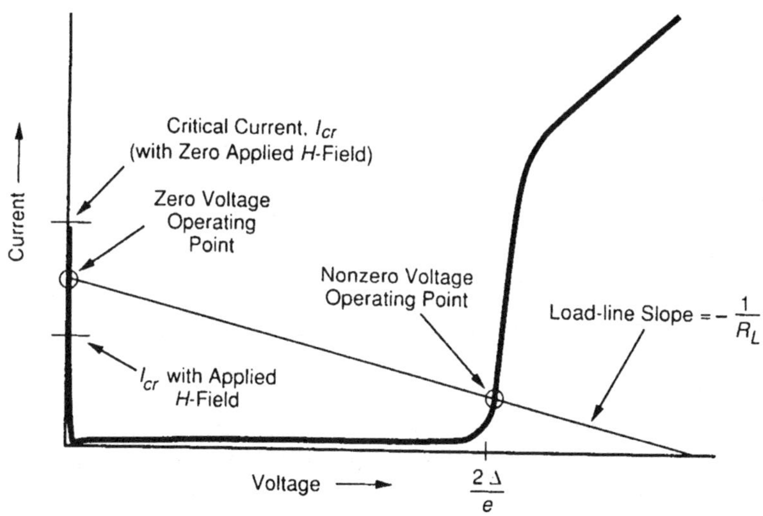

FIGURE 8.13. Current, I/voltage, V characteristic curve of a Josephson junction.

function point between the Josephson state ($V \approx 0$) and the Giaever state ($V = V_g$).

8.4.2 Superconducting Quantum Interference Device (SQUID)

The essential element of a SQUID is a Josephson junction placed in series with current path around the hole containing the flux. This junction has a layered structure usually of S-I-S configuration, see Figure 8.14(a). An alternative construction of the basic SQUID element is based on weak links. A typical weak link consists of a very narrow bridge of superconducting material that provides a conduction path between two larger parts of superconducting material, see Figure 8.14(b); similar I-V characteristics to Josephson junctions are observed, see Figure 8.14(c). Thin-film and photolithographic techniques are dedicated to the fabrication of SQUID components.

Two types of SQUID exist: single and double weak-link SQUID, called also ac and dc SQUID, respectively. A schematic illustration of both types of SQUID are shown in Figure 8.15. The essential features of SQUID devices must be biased, so that small changes in flux will result in exceeding the weak-link critical current, causing a voltage difference at the interface. Then, when the weak-link voltage is detected, external feedback circuitry adjusts flux (or current) to reverse the whole condition. The feedback response is that which gives the SQUID output. The sensitivity of SQUIDs is very high; the measuring flux changes are of the order of 10^{-4} Φ_o(Φ_o = flux quantum = 2.07×10^{-15} Wb).

8.4.3 Microwave Antennas

The new ceramic superconductors can be used in antenna applications. HTS films, with very low surface resistance, possess excellent performance in microwave passive devices, compared with normal conductors. Many superconductive microwave devices need to work in strong dc-magnetic fields. When magnetic fields, exceeding the lower critical field, namely H_{c1}, are applied, the HTS will be in the mixed state and the vortex lattice will form. The rf-field, fed into the superconducting devices, interacts with the vortex lattice, which will affect the microwave surface impedance of the HTS.

A typical construction of a HTS microstrip antenna (YBCO/$LaAlO_3$) is shown in Figure 8.2(e). The HTS film can be deposited on appropriate substrates by various techniques, whereas the antenna patterning can be performed by photolithography and etching techniques [25]. Microwave antennas can also be fabricated by wire-mounting techniques using bulk polycrystalline HTS materials.

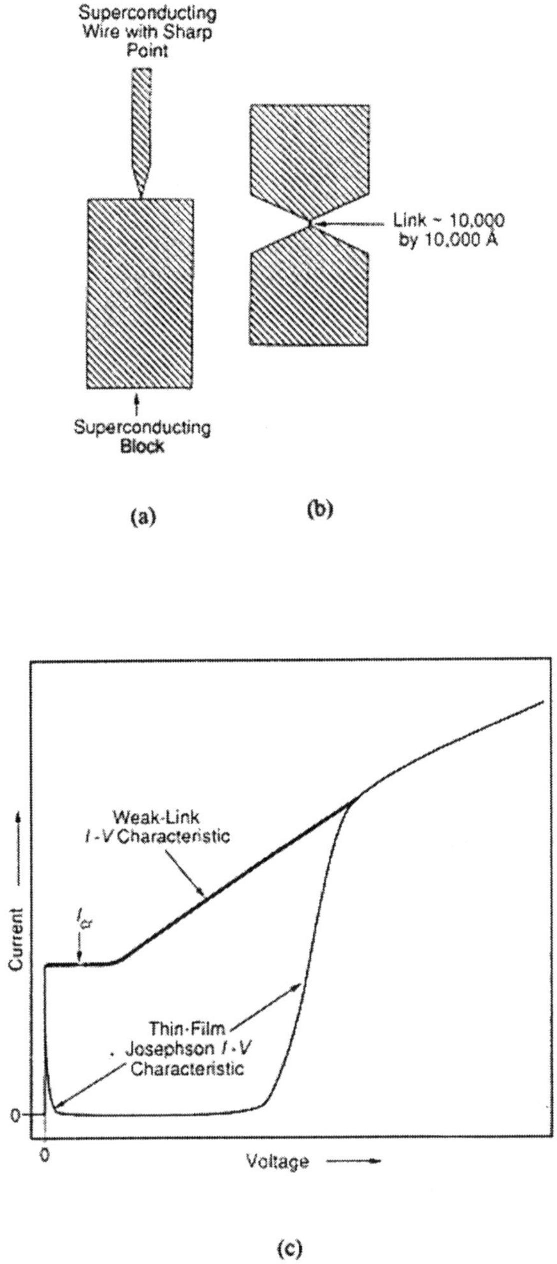

FIGURE 8.14. (a) S-I-S component used for SQUID applications. (b) Typical weak-link element used for SQUID application. (c) I-V curves for Josephson junctions and weak links (from Reference [1]).

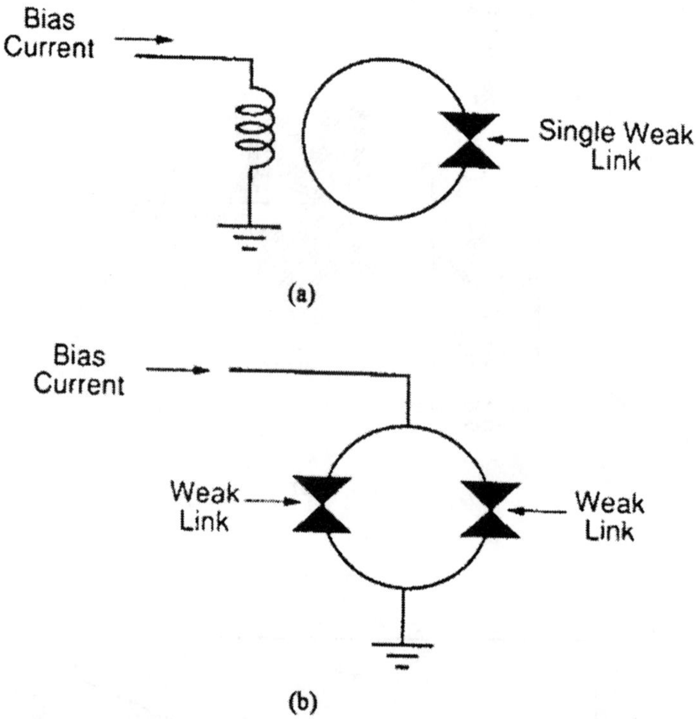

FIGURE 8.15. (a) Single and (b) double weak-link SQUID.

8.4.4 Radiation Detectors

Superconductive films can be used for detection of microwave and IR radiation. Superconducting radiation detectors are advantageous, because:

- less power is necessary for the function of the detector
- their response is rapid
- band width is very broad, extending to far IR

The detection techniques are generally classified as thermal (bolometric) and nonthermal. In the bolometric technique, the temperature of the sensor must be increased by the absorption of radiation, causing an abrupt increase of the resistivity, when the sensor is operating near T_c. In the latter technique, the function of the sensor depends on the nonlinearity in the *V-I* characteristic curve of the film. In the nonthermal mechanism, the function of the detector is based on the creation of quasi-particle pairs induced by the incident photon. The actual measurement constitutes of a reduced gap energy, which is a modification of the Josephson *V-I* curves [26]. Figure 8.16 shows a

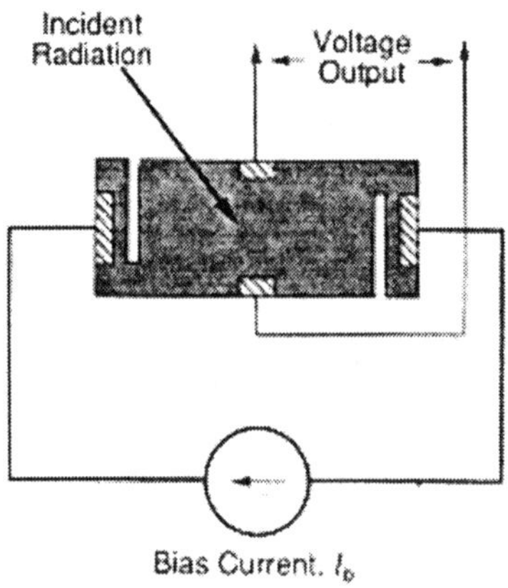

FIGURE 8.16. A schematic illustration of a HTS radiation detector (from Reference [26]).

schematic representation of a superconducting thin-film bolometer for far IR microwave detection, see Reference [27].

8.4.5 Computer Applications

Josephson junctions are of great importance in novel superconducting computer technology. These junctions can be used as switching elements and, therefore, they can be connected to behave as gates and memory elements in computer circuits. The basic principle is that the junction voltage function point can be switched from 0 to 2–3 mV by changing either the interface current or either the local magnetic field. The advantages of the Josephson junctions in computer circuits are the very high speed and low dissipation. Gate switching times of the order of 10 ps can be achieved, whereas their area is up to 1.5–2.0 μm^2. Switching power dissipation is typically of the order of a few microwatts per gate at maximum speed, which is three orders of magnitude lower than that of the best semiconducting gate.

COMPUTER GATES

Computer gates can be divided into two categories: direct-coupled and magnetic or induction coupled. In the direct coupled gate the junction is switched by the injection of an additional current into the gate. The combination of junction and load resistance can be constructed with very low inductance, resulting in very fast switching. The induction coupled gate uses a SQUID configuration, where the introduction of a very small magnetic field is produced by a pulsed current in one or more control lines near the junction [28].

COMPUTER MEMORY ELEMENTS

In conventional computers, the entire random access memory (RAM) is usually of one type. In Josephson superconducting computers, memory is typically segregated into two sections. The fastest section is also the smallest (in bytes of storage); this is referred to as cache memory and is used to work directly with the central processing unit (CPU). With the cache memory present to handle the fast interactions with the CPU, the main RAM, which will usually consist of more than 10 megabytes, may be slower, trading off speed for lower power dissipation. Such memory cells provide an access time on the order of 10 ns and consume no power except during read and write cycles [29].

8.5 REFERENCES

1. Doss J. D. (1989), *Engineer's Guide for High Temperature Superconductivity,* John Wiley & Sons, New York.
2. Vajda I., Mamalis A. G., Szalay A. and Göbl N. (1996), "Design principles of a high-temperature superconducting rotating electrical machine," *Proc. Int. Conference on Electrical Machines, ICEM 96,* Vigo, Spain.
3. Vajda I., Mamalis A. G. and Szalay A. (1996), "Design and construction of an HTSC synchronous machine with permanent magnet excitation," *Proc. Int. Cryogenic Engineering Conference, ICEC 16/ICMC,* Kitakyushu, Japan.
4. Mamalis A. G., Vajda I., Szalay A., Kotsis I. and Pantazopoulos G. (1997), "Some small-scale high-temperature superconducting models for applications in electricity and transportation," *Superlatt. Microstruct.* 21 (Suppl. A), 251.
5. Mamalis A. G., Kotsis I., Vajda I, Szalay A. and Pantazopoulos G. (1996), "High-T_c ceramic superconductors for electrical rotating machines: from fabrication to application," *Proc. 10th Anniversary HTSC Workshop on Physics, Materials and Applications,* Houston, Texas, 631.
6. Mamalis A. G., Szalay A., Vajda I., Enisz M. and Palotás B. (1995), "A novel monolith for electrical applications: silver sheathed bulk ceramic superconductor with silver contacts," *Proc. 4th Euro-Ceramics Conf., ECERS,* Barone A., Fiorani D., Tampieri A., eds. Gruppo Editoriale Faenza Editrice S.p.A., Italy, 7, 217.
7. Szalay A., Vajda I, Mamalis A. G., Kotsis I., Bankuti J., Göbl N. and Palotás B. (1996), "A novel metal/ceramic monolith: silver sheathed bulk high-T_c superconductor with silver contacts," *Proc. 2nd Japan-Central Europe Joint Workshop on Modelling of Materials and Combustion,* Budapest, Hungary, 155.
8. Mamalis A. G., Vajda I., Mohacsi L., Kotsis I. and Szalay A. (1995), "From fabrication to application: ceramic superconductors for use in magnetically levitated bearings," *Proc. 4th Euro-Ceramics Conf., ECERS,* Barone A., Fiorani D., Tampieri A., eds. Gruppo Editoriale Faenza Editrice S.p.A., Italy, 7, 167.
9. Mamalis A. G., Szalay A., Pantelis D. and Pantazopoulos G. (1995), "Fabrication of thick layered superconductive ceramic (Bi-Pb-Sr-Ca-Cu-O)/metal composite strips by explosive cladding and rolling," *J. Mater. Proc. Technol.* 51, 255.

10. Kotsis I. and Enisz M. (1992), "Effect of alkali fluoride additives on the properties of superconductors of the system Y-Ba-Cu-O," *Proc 8th SIMCER,* Rimini, Italy.

11. Kotsis I., Enisz M., Oravetz D. and Szalay A. (1995), "Effect of porosity on properties of explosively compacted high-T_c superconductors," *Hungarian J. Industrial Chemistry Veszprém* 23, 69

12. Singh J. B., Joo J., Singh D., Warzynski T. and Poeppel R. B. (1993), "Effects of silver additions on thermal shock and delayed fracture of $YBa_2Cu_3O_{7-\delta}$," *J. Mate. Res.* 8, 1226.

13. Mamalis A. G., Szalay A., Pantelis D. I., Pantazopoulos G., Kotsis I. and Enisz M. (1996), "Fabrication of multi-layered steel/ superconductive ceramic (Y-Ba-K-Cu-O)/silver rods by explosive powder compaction and extrusion," *J. Mate. Proc. Technol.* 57, 155.

14. Salama K. and Lee D.F. (1994), "Progress in melt texturing of $YBa_2Cu_3O_x$ superconductors," *Supercond. Sci. Technol.* 7, 177.

15. Bánkuti J., Vajda I., Mohácsi L., Szalay A., Kotsis I. and Enisz M. (1996), "Design and construction of a small-scale model of a high-temperature superconducting magnetically levitated vehicle," *IEEE Trans. on Magnetics* 32, 2288.

16. Mamalis A. G., Vajda I., Ioannidis M. B., Monacsi L. and Szalay A. (1995), "Finite difference modelling of magnetic characteristics of high-T_c superconductive ceramics," *Superconductor Sci. Technol.* 8, 579.

17. Kirschner I., Leppävuori S., Laiho R., Altfeder I., Dódony I., Uusimäki A., Porjesz T., Hagberg J., Kovács G., Lähderanta E., Volodin A. and Zsolt G. (1991), "Electrical and magnetic properties Bi(Pb)-Sr-Ca-Cu-O thick films annealed variously," *Z. Phys.* B85, 175.

18. Ratnajevan S. and Hole H. (1989), *Computer-Aided Analysis and Design of Electromagnetic Devices,* Elsevier, New York.

19. Malozemoff A. P., Carter W., Fleshler S., Fritzemeier L., Li Q., Masur L., Miles P., Parker D., Parrella R., Podtburg E., Riley G.N., Jr., Rupich M., Scudiere J. and Zhang W. (1998), "HTS wire at commercial performance levels," *Proc. Applied Superconductivity Conference,* Palm Desert, California.

20. Manlief M. D., Picard H., Snitchler G., Ghosh A. K., Harrison M., Sampson W. B., Wanderer P., Dugan G., Rogers J., Temnykh A., Brandt J. S., Limon P. J. and Sood R. C. (1998), "The performance of Bi-Sr-Ca-Cu-O superconducting quadrupole coils," *Proc. Applied Superconductivity Conference,* Palm Desert, California.

21. Lewis, J. A., Childs R. A. (1999), "Heat treating a superconducting coil for a fusion reactor," *Advanced Materials & Processes* 1, 67.

22. Buchhold T. A. (1975), "Superconductive machinery," in *Applied Superconductivity,* V. L. Newhouse, ed., John Wiley & Sons, New York.

23. Singh S. K., Scherbatth D. W., Ortoli E. S., Repp J. R., Christianson O. R., Parker J. H., Carr J. W., Jr., Rhoads J. L., Oberly C. E. and Gamble B. (1998), "Conceptual design of a high temperature superconducting generator," *Proc. Applied Superconductivity Conference,* Palm Desert, California.

24. Nakajima A., Terui F., Sawada H. and Kohno T. (1996), "High-T_c superconductors for aerospace applications," *Proc. 2nd Japan-Central Europe Joint Workshop on Modelling of Materials and Combustion,* Budapest, Hungary.

25. Meihong Z., Bisong C., Na L., Aisheng H., Xuexia Z., Yonggang Z., Daijie D., Yujie W., Yueliang Z., Huibin L., Dafu C., Changlian Y., Kunlun L., Yanju W., Tiejun L. and Binglin G. (1996), "Effects of dc magnetic field upon the properties of a HTS microstrip antenna at 4.2 K," *Appl. Supercond.* 4, 241.

26. Talvacchio J., Forrester M. G. and Braginski A. I. (1989), "Photodetection with high-T_c superconducting films," in *Science and Technology of Thin-Film Superconductors,* R. McConel, S.A. Wolf, eds., Plenum Press, New York.

27. Rose K., Bertin C. L. and Katz R. M. (1975), "Radiation detectors," in *Applied Superconductivity, Vol. 1,* V. L. Newhaouse, ed. Academic Press, New York, 267.

28. Gheewala T. R. (1980), "Josephson logic devices and circuits," *IEEE Trans. Electron Devices* ED-27, 1857.

29. Van Duzer T. (1980), "Josephson digital devices and circuits," *IEEE Trans. Microwave Theory & Technol.* MTT-28, 490.

CHAPTER 9

Future Perspectives of High-T_c Superconductivity

9.1 NOVEL MATERIALS AND APPLICATIONS

The tendency in high-T_c superconductivity nowadays is toward new superconductive systems and their applications. A novel outline on these new materials is given below; extended reports on these topics can be found in Reference [1].

9.1.1 The "BaCaCu" Oxides and Oxycarbonates

The high-pressure compounds in the system Ba-Ca-Cu-O, first reported as cuprates, with a critical temperature, T_c, ranging from 60 to 117 K, are in fact oxycarbonates due to the contamination from the high-pressure graphite film used. The structure of these compounds consists of infinite CuO_2 layers interconnected with rows of carbonate and copper as shown in Figure 9.1(a) for the $(Cu_{1-y}Cu_y)Ba_2Ca_2Cu_3O_{8+d}$ compound, see Reference [2]. Very similar superconductors, with a T_c up to 110 K, have been synthesized in the form of thin films. In the latter, the CuO_2 layers are interrelated with layers fully occupied by carbonates, see Figure 9.1(b). In a recent investigation, see Reference [4], a pure cuprate $Ba_2Ca_{1-x}Cu_xCa_2Cu_3O_8$, without carbonate, was synthesized at high pressure, with a T_c of 105 K. The latter exhibits the 1223-type structure, in which $[Ba_2Ca_{1-x}Cu_xO_3]$ layers play the role of hole reservoir.

9.1.2 $A_{1-x}B_xCuO_2$ Compounds

The homologous series of layered $Ca_{n-1}Cu_nO_{2n-2}$ with large values of n, which form the center part of Bi-22(n-1)n or Tl-22(n-1)n compounds, is a

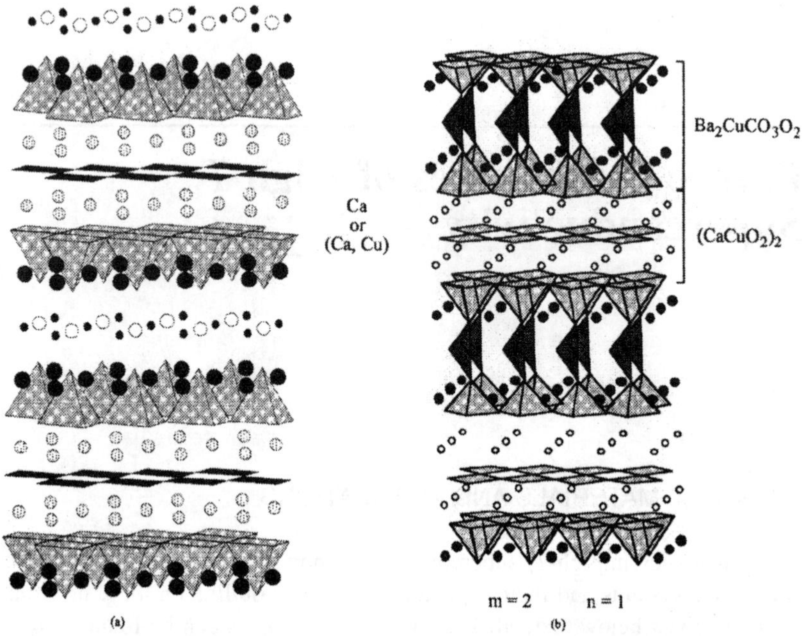

FIGURE 9.1. Lattice models for the $(Cu_{1-y}Cu_y)Ba_2Ca_2Cu_3O_{8-d}$ compound (from Reference [2]).

newly developed member of the high-temperature superconductor family, see Reference [5].

Significant members of the $A_{1-x}B_xCuO_2$ family are the $(Ca_{0.85}Sr_{0.15})CuO_2$ compound, which is an infinite layered compound but not superconductive, the electron-doped $Sr_{1-y}Nd_yCuO_2$ compound, which is synthesized under 2.4 GPa and possesses a T_c = 40 K and the 110 K superconductor $(Sr_{1-x}Ca_x)_{0.9}$, prepared under 6 GPa. The highest T_c will be performed by the hole-doped infinite layered or by the 1T′-structure cuprate. The structure, synthesis and the related properties of these compounds are still under progress.

9.1.3 Sr-Cu-O Compounds

Superconductivity up to 90–100 K was reported for Sr-Cu-O compounds synthesized under high pressures. The main members of the Sr-Cu-O system are the $SrCuO_2$, $Sr_2Cu_{3+\delta}$, $Sr_3Cu_2O_{5+\delta}$ and $Sr_4Cu_3O_{7+\delta}$ compounds. The $SrCuO_2$ and $Sr_2CuO_{3+\delta}$ compounds are not superconductive, based on the Cu-O bond length. The superconducting compounds in the Sr-Cu-O system can be synthesized in a multianvil press at 800–1100°C temperature range under pressure condition of 6–8 Gpa, see Reference [5]. The superconductivity is due to the $Sr_3Cu_2O_{5+\delta}$ phase content, whereas the T_c of the Sr-Cu-O system increases with $Sr_3Cu_2O_{5+\delta}$ content from 70 to 100K.

9.1.4 $Sr_2YRu_{1-x}Cu_xO_6$ Compounds

The common features of the high-temperature superconductors are their parent compounds being antiferromagnetic (AF) insulators, consisting of layered structure based on a perovskite lattice with CuO_2-square planes. Proper doping of an AF insulating compound, without CuO_2-planes, may induce superconductivity. The double perovskite compound A_2YRuO_6, where A is a divalent earth element, is a well-known AF insulating compound. The recent systematic study of the structural stability of the transition-metal doped $Sr_2Y(Cu_{3-x}M_x)O_y$ has led to the discovery of the $Sr_2Y(Ru_{1-x}Cu_x)O_6$ superconductor, see Reference [7]. The onset of the superconducting transition, determined by resistivity and magnetic measurements, occurs at temperatures varying from 30 to 85 K, depending on the Cu-content and the processing conditions.

9.1.5 Superconducting Oxycarbonates

The copper oxycarbonates form a very promising family of high-T_c superconductors with closely related structures. The synthesis of these materials is based on the infinite layer structure, considered as a host lattice for the formation of the oxycarbonate $Sr_2CuO_2CO_3$(S_2CC) and on the combination of the S_2CC structure with the $Tl_{0.5}Pb_{0.5}Sr_2CuO_5$ (1201) and $Bi_2Sr_2CuO_6$ (2201) type structures [8].

The formation of this family of oxycarbonate compounds is based on intergrowth and crystallographic shearing mechanisms; for more details see Section 3.5 and the related references. The mechanisms of superconductivity in these phases, such as the role of CO_3 groups and of the Cu valence, are not well understood yet, but the high pinning capability and the possible 110 K superconductivity are very promising for electronic applications.

9.1.6 Some Novel Applications

MAGNETOCARDIOGRAPHY USING HIGH-T_C SQUIDS

The scope of this application is to develop a magnetocardiography (MCG) system that is capable to measure magnetocardiograms in an unshielded environment. This can be succeeded by using HTS SQUIDS, see Reference [9]. The additional noise exhibited by SQUID can be eliminated by zero-field cooling.

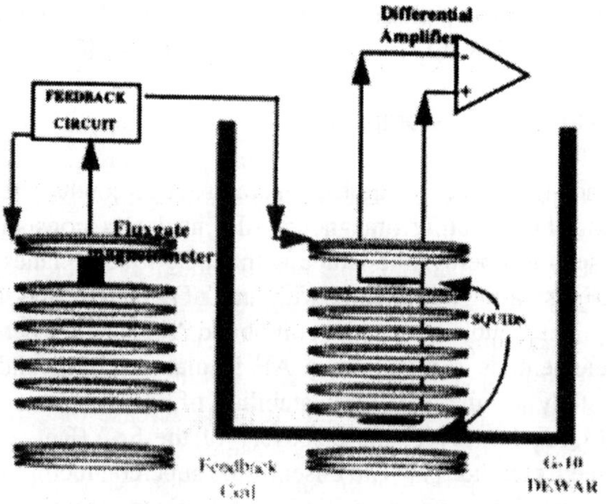

FIGURE 9.2. Schematic illustration of the MCG system (from Reference [9]).

Figure 9.2 shows one of those measuring systems. A fluxgate magnetometer is used to track the earth's field and its low frequency (LF) fluctuations, as well as the powerline noise. It enables also to the zero-field cooling of the HTS SQUID and the passive shields. Note that localized HTS shields are corporated to selectively shield out the noise field without any affection of the interested dipole source fields. The entire setup is supported by a nonmetallic shell to eliminate eddy current noise.

DYE/SUPERCONDUCTOR ELECTRONIC SENSORS

This dye/superconductor assembly can be used as a basic component of a wavelength-selective bolometric sensor. The composite detector uses a molecular organic chromophore material, with wavelength-dependent absorbing properties. The light-harvesting antenna layer is shown to absorb certain wavelengths, and it transfers this light energy to the superconducting sensing element. The maximum sensitivity of the detector can be tuned to various wavelengths of light in the visible and near IR spectral regions. A classic method to fabricate this organic/superconductor composite is composed of the following steps, see Reference [10]:

- pulsed-laser deposition of the superconducting film
- laser patterning of the superconducting layer to form a microbridge
- thermal evaporation of the organic dye over the previously patterned region of the HTS thin film

9.2 SUPERCONDUCTIVITY AT ROOM TEMPERATURE: REALITY OR DREAM?

High-temperature superconductivity has allowed many opportunities for the realization of practical applications working in a more realistic manner determined by the coolant used. The first breakthrough is made by using liquid nitrogen for the 90 K-class superconducting materials. The next targets include the discovery of materials that may be practical superconductors by using Freon (CF_4) up to 148 K, dry ice up to 198 K and room temperature up to 298 K as the cooling environment [5].

The transition temperature at 134 K, e.g., at atmospheric pressure, or at 164 K using higher pressures is due to the decrease of CuO_2-interplanar spacing, which is equivalent to the increase of CuO_2 layers (n) per unit-cell volume followed by proper doping. It has been shown that the critical temperature increases by increasing the number of holes, *h,* up to 3 or 4, with the maximum T_c corresponding to carrier concentration density, *c* equal to 0.16 holes/Cu-ion. The drop in T_c may be mainly attributed to the possible depletion of the *c*-density in the interior layers in a unit cell, see Reference [5].

Optimization of carrier concentration density can be achieved by doping. Note, however, that sometimes doping conditions may lead to structural instabilities followed by decomposition of the original $n > 3$ members. This is consistent with high values of Coulomb repulsion, which arises from the development of excessive charge as the *n* increases. Appropriate doping, reaching optimal values of *n* and *c* and avoiding doping instability, may lead to developing materials with higher T_c. It may be also noted that cuprate superconductors possess undesirable instabilities, such as structural collapse, magnetic ordering, formation of charge density waves or spin density waves, leading, in many cases, to erratic or destroyed superconductivity. These catastrophic characteristics are mainly based on the superconductivity mechanism that prevailed in the cuprate ceramics. Novel materials, presented above with possible new superconductivity mechanisms, are promising candidates for higher critical temperature applications.

In addition, the better and deeper understanding of "intimate" superconductivity mechanisms and the overall theory around them, including the effect of crystal structure/doping, may encourage the creation of new materials with considerably higher T_c, enhancing also the structural stability and the short coherence length. When this dream T_c increase and how close to the room temperature is going to be materialized is the future perspective of high-T_c superconductivity.

9.3 REFERENCES

1. *Proceedings of the 10th Anniversary HTS Workshop on Physics, Materials and Applications, Houston*, Texas (1996), B. Batlogg, C. W. Chu, W. K. Chu, D. U. Gubser and K. A. Müller, eds. World Scientific Publishing, Singapore.
2. Chaillout C. and Marezio M. (1996), "Enhancement of T_c for $(Cu_{0.5}C_{0.5})Ba_2Ca_2Cu_3O_{8.5+\delta}$ from 67 to 120 K by heat treatments," *Proc. 10th Anniversary HTS Workshop on Physics, Materials and Applications,* Houston, Texas, B. Batlogg, C. W. Chu, W. K. Chu, D. U. Gubser and K. A. Müller, eds. World Scientific Publishing, Singapore, 113.
3. Allen J. L., Mercey B., Prellier W., Hamet J. F., Hervieu M. and Raveau B. (1995), "The first superconducting thin-film oxycarbonates : coherent intergrowth of several members of a new structural series $(CaCuO_2)_m(Ba_2Cu_2O_2CO_3)_n$," *Physica C* 241 (1995), 158.
4. Wu N. L., Du Z. L., Xue Y. Y., Rusakova I., Ross D. K., Gao L., Cao Y., Sun Y. Y., Chu C. W., Hervieu M. and Raveau B. (1998), "Observation and identification of a 107 K superconducting family in the $CaBa_2Ca_2Cu_3O_{9-\delta}$ system," *Physica C* (in press).
5. Chu C. W. (1996), "Superconductivity above 90K and beyond," *Proc. 10th Anniversary HTS Workshop on Physics, Materials and Applications,* Houston, Texas, B. Batlogg, C. W. Chu, W. K. Chu, D. U. Gubser and K. A. Müller, eds. World Scientific Publishing, Singapore, 17.
6. Du Z. L., Cao Y., Sun Y. Y., Gao L., Xue Y. Y. and Chu C. W. (1996), "Superconducting phases in the Sr-Cu-O system," *Proc. 10th Anniversary HTS Workshop on Physics, Materials and Applications,* Houston, Texas, B. Batlogg, C. W. Chu, W. K. Chu, D. U. Gubser and K. A. Müller, eds. World Scientific Publishing, Singapore, 123.
7. Ling D. C., Sheen S. R., Tai S. Y., Tseng J. L., Wu M. K., Chen T. Y. and Chien F. Z. (1996), "Superconductivity in $Sr_2YRu_{1-x}Cu_xO_6$ system," *Proc. 10th Anniversary HTS Workshop on Physics, Materials and Applications,* Houston, Texas, B. Batlogg, C. W. Chu, W. K. Chu, D. U. Gubser and K. A. Müller, eds. World Scientific Publishing, Singapore, 129.
8. Raveau B., Michel C. and Hervieu M. (1996), "Generation of high-T_c superconductivity oxycarbonates: from the ceramics to the thin films," *Proc. 10th Anniversary HTS Workshop on Physics, Materials and Applications,* Houston, Texas, B. Batlogg, C. W. Chu, W. K. Chu, D. U. Gubser and K. A. Müller, eds. World Scientific Publishing, Singapore, 117.
9. Miller J. H., Jr., Tralshawala N., Claycomb J. R. and Xu J. H. (1996), "Magnetocardiography in a magnetically noisy environment using high-T_c SQUIDS," *Proc. 10th Anniversary HTS Workshop on Physics, Materials and Applications, Houston, Texas,* B. Batlogg, C. W. Chu, W. K. Chu, D. U. Gubser and K. A. Müller, eds. World Scientific Publishing, Singapore, 639.
10. McDevitt J. T., Jurbergs D. C., Savoy S. M., Eames S. J. and Zhao J. (1996), "Electronic eyes based on dye/superconductor assemblies," *Proc. 10th Anniversary HTS Workshop on Physics, Materials and Applications,* Houston, Texas, B. Batlogg, C. W. Chu, W. K. Chu, D. U. Gubser and K. A. Müller, eds. World Scientific Publishing, Singapore, 613.

AUTHOR INDEX

SUBJECT INDEX